An Introduction to the Technique of Formative
Processes in Set Theory

Domenico Cantone · Pietro Ursino

An Introduction
to the Technique
of Formative Processes
in Set Theory

 Springer

Domenico Cantone
Dipartimento di Matematica e Informatica
Università degli Studi di Catania
Catania
Italy

Pietro Ursino
Dipartimento di Fisica e Matematica
Università degli Studi dell'Insubria
Como
Italy

ISBN 978-3-319-89283-2   ISBN 978-3-319-74778-1   (eBook)
https://doi.org/10.1007/978-3-319-74778-1

Printed on acid-free paper

This Springer imprint is published by the registered company Springer
International Publishing AG part of Springer Nature
The registered company address is: Gewerbestrasse 11, 6330 Cham, Switzerland

To Maria Pia and Riccardo,
and to the memory of my beloved parents

To Melita and Miranda

Preface

The decision problem in set theory has been explored intensively in the last decades, mainly with the following goals:

- mechanical formalization of mathematics with a proof verifier based on the set-theoretic formalism [FOS80, OS02, COSU03, OCPS06, SCO11],

- enhancement of the services offered by high-level specification languages including sets and functions among their primitives [SDDS86, KP95],

- handling set constraints [DPPR98, RB15, CRF15] in an advanced declarative programming setting.

This research has generated a large body of results, giving rise to the field of *Computable Set Theory* [CFO89, CF95, COP01, SCO11].

In the design of decision procedures in set theory, one needs inevitably to face the problem of manipulating collections of sets, subject to certain relationships among them that must be maintained invariant. In fact, for a given fragment $\mathfrak{F}$ of set theory of which one is investigating the satisfiability problem, a typical approach consists of finding a *uniform method* that simplifies the satisfying assignments of any given formula in $\mathfrak{F}$, in such a way that some of their representatives—and, possibly, some additional finite structures attached to them—can be confined into a finite collection of candidate assignments that can be effectively generated and tested for satisfaction. As we shall see, this amounts to proving that $\mathfrak{F}$ enjoys a *small model property* or a *small witness-model property*.

This book provides an introduction to the most advanced of these simplification methods developed in computable set theory, namely, the *technique of formative processes*. Each set carries in its extension a complete 'construction plan', which can be conveniently unraveled and manipulated according to one's needs. In the case of a finite number of sets $\mathcal{F}$, which collectively may form a satisfying assignment for a given set-theoretic formula, it proves advantageous to represent $\mathcal{F}$ through its Venn partition, or some of its extensions, rather than by the mere collection of its members. The 'construction plan' $\mathcal{H}_{\mathcal{F}}$ of such an extended Venn partition Σ of $\mathcal{F}$ is a *formative process*. By appropriately manipulating $\mathcal{H}_{\mathcal{F}}$, one can produce other formative processes whose induced set collections $\mathcal{F}'$ enjoy the same properties as $\mathcal{F}$, plus possibly additional ones (e.g., rank boundedness or infinite cardinality).

In the mid-1980s, a similar technique began to emerge in the process of solving the satisfiability problem for the fragments of quantifier-free formulae of set theory called MLSP (viz., *Multi-Level Syllogistic with the Powerset operator*; see [CFS85]) and MLSSP (see [Can87] and [Can91]). Specifically, MLSP-formulae are the propositional combinations of atoms of the following types

$$x = y \cup z, \qquad x = y \cap z, \qquad x = y \setminus z, \qquad x \in y, \qquad x = \mathcal{P}(y),$$

(where x, y, z stand for set variables, and $\mathcal{P}(\cdot)$ is the powerset operator), whereas MLSSP-formulae are obtained by extending the endowment of MLSP with atoms of the form $x = \{y\}$, involving the singleton operator. This technique was then restated in [CU97, COU02], in much like the current terms, and further refined in [Urs05, CU14], where it was applied to solve the satisfiability problem for the extension MLSSPF of MLSSP with atomic formulae of the form $Finite(x)$, intended to express that x has a finite cardinality. However, unlike MLSSP-formulae, some formulae in MLSSPF are satisfied by infinite assignments only. Hence, the fragment MLSSPF cannot enjoy the *classical* small model property. Instead, the decidability of MLSSPF can be obtained by proving a *small witness-model property*, using the technique of formative processes. Specifically, given an MLSSPF-formula Φ admitting only infinite satisfying assignments, one can show that the associated formula Φ^-, obtained by dropping from Φ all literals involving the predicate $Finite(\cdot)$, admits a small satisfying assignment whose formative process can be 'pumped' into a new infinite formative process.

An approach strictly related to the technique of formative processes was also undertaken by Omodeo and Policriti in [OP10, OP12] for the satisfaction problem of the Bernays-Schönfinkel-Ramsey (BSR) class of set-theoretic statements (namely, the class of prenex statements, with a prefix of the form $\exists^*\forall^*$, involving the predicate symbols of equality '=' and membership '∈' only). Since BSR-statements with a prefix of the form $\exists\exists\forall\forall$ can express infinite sets, as proved in [PP88], the decidability result for the BSR-class has been established by proving for it a small witness-model property. Specifically, it has been shown that any satisfiable BSR-formula Φ admits a non-necessarily finite model that can be represented by a finite graph of *bounded size* with special nodes, called *rotors*, standing for denumerably infinite families of *well-behaved* sets.

In the form in which they were originally conceived, several decision algorithms in computable set theory are often hard to extend with new set-theoretic constructs that were not built into them from the outset. We expect that a full understanding of the technique of formative processes can enhance its applicability, simplifying combinations of decision procedures, and possibly giving a new research boost to the fascinating area of computable set theory. In fact, the consolidation of the known part of computable set theory is essential not only to promote new discoveries on decidability, but also to convert the theoretical results into technological advances in the field of automated reasoning. For instance, even the most basic layer of automated set reasoning, the so-called *multi-level syllogistic* (MLS), has benefited from being revisited under a tableaux-based approach, which renders its implementation far more efficient (see [CZ00, CZ99] and [COP01, Chapter 14]).

We hope that the present book may contribute to a renewed interest in computable set theory and, specifically, in the related technique of formative processes. We expect that the technique of formative processes might be an essential tool in the solution of some long-standing open problems in the area, such as the satisfiability problem for the extension MLSC of multi-level syllogistic with the Cartesian product. This appears to be particularly relevant insofar as the satisfiability problem for MLSC can be seen, in some sense, as the set-theoretic counterpart of the well-known Hilbert's tenth problem.

Content: The book is divided into two parts. The first part is preliminary and consists of three chapters. Chapter 1 introduces the basic set-theoretic terminology and properties used throughout the rest of the book. Chapter 2 is an introduction to the decision problem in set theory. After presenting the syntax and semantics of a comprehensive theory that encompasses all the fragments of set theory considered in the book, the novel notion of *satisfiability by partitions* is introduced. The relations of *simulation* and *imitation* among partitions, which maintain invariant the collection of formulae satisfied by them, are then examined, and a simple case study—the theory BST—is explored in detail. It is proved, in particular, that BST has an NP-complete decision problem. Subsequently, various expressiveness results are obtained and then used to prove two negative results about decidability and *rank dichotomicity*. The latter is a novel concept, strictly related to the small model property. Chapter 3 is devoted to *formative processes*. Preliminarily, formative processes, their supports—*syllogistic boards*—and their applications to the decision problem of MLSP-like fragments are illustrated by means of worked-out examples and pictures. Then all these notions are rigorously presented along with some of their properties. Finally, the so-called *shadowing relationship* among formative processes and its properties are discussed, proving that the final partitions of two shadowing formative processes imitate each other.

The second part of the book consists of two chapters, which are devoted to applications of the technique of formative processes to the decision problems for the theories MLSSP and MLSSPF, respectively. Specifically, given a partition Σ and a formative process $\mathcal{H}$ for it, Chapter 4 presents a technique for extracting the subsequence $S_{\mathcal{H}}$ of the *salient steps* of $\mathcal{H}$. The subsequence $S_{\mathcal{H}}$ enjoys the following two properties: (i) its length is bounded by a computable function of the size of the partition Σ, and (ii) there exists a formative process $\widehat{\mathcal{H}}$ whose sequence of steps is $S_{\mathcal{H}}$ and is such that $\widehat{\mathcal{H}}$ is a shadow process of $\mathcal{H}$. This amounts to showing that MLSSP enjoys the small model property, thereby proving its decidability. In addition, the shadow process $\widehat{\mathcal{H}}$ preserves certain structural properties of $\mathcal{H}$. This fact is exploited in Chapter 5 to prove the existence of suitable infinite shadow processes of $\mathcal{H}$ (*pumping mechanism*). Let Φ be an MLSSPF-formula and, again, let Φ^- be the MLSSP-formula obtained by dropping from Φ all finiteness literals. Then it turns out that one can generate effectively all (descriptions of) infinite processes that are in the shadowing relationship with the formative process of some satisfying assignment for Φ^- (*small witness-model property*). As proved in Chapter 5, the latter fact yields the decidability of MLSSPF.

All chapters (especially in the first part) contain a number of exercises to help the reader become familiar with the new concepts and techniques. Some exercises simply ask the reader to provide or complete omitted proofs; others are devoted to deepening the comprehension of the topics discussed in the main text. The reader is strongly encouraged to work them out for a full understanding of the concepts and techniques of the book.

Besides some "mathematical maturity," the reader is assumed to have familiarity with the basic set-theoretic apparatus, as well as with some elementary notions from mathematical logic and computability.

Finally, to enhance readability, each proof is terminated by the empty square symbol $\Box$, whereas all examples and remarks are closed by the full square symbol $\blacksquare$.

Acknowledgments

We are grateful to Eugenio Omodeo and Alberto Policriti for some initial and very enjoyable conversations in the late 1990s, when the technique of formative processes began to take shape in its current form. We are also thankful to Alfio Giarlotta, who provided very helpful comments on the first part of the book and was kind enough to draw one of the figures, and to Ronan Nugent, Senior Editor at Springer Computer Science, for his continuous assistance and support. The second author is also indebted to Josie Nacci for her kind help in English proofreading Chapters 4 and 5.

Contents

Part I

Theoretical Apparatus

Chapter 1

Basics of Set Theory

We briefly recall some basic set-theoretic terminology which will be used throughout the book.

Our considerations will take place in a naive set theory that could be formalizable in the standard axiom system ZFC, developed by Zermelo, Fraenkel, Skolem, and von Neumann (see [Jec78]). In particular, we shall refer to the von Neumann standard cumulative hierarchy of sets, a very specific 'model' of ZFC, and will assume the *Axiom of Regularity* (or *Axiom of Foundation*) as the unique *less naive* principle:

AXIOM OF REGULARITY:
Every non-null set X contains a member x that is disjoint from X, i.e., such that $x \cap X = \emptyset$.

The Axiom of Regularity prevents the formation of cycles of memberships of the form

$$x_0 \in x_1 \in \cdots \in x_n \in x_0$$

and, more in general, of infinite descending chains of memberships

$$x_0 \ni x_1 \ni \cdots \ni x_n \ni \cdots . \tag{1.1}$$

In fact, if there were an infinite descending chain of memberships of the form (1.1), then the set $X = \{x_0, x_1, \ldots\}$ of all the elements participating in such a chain would violate the Axiom of Regularity, as each member of X would intersect X.[1]

Other axioms of set theory that we shall sporadically mention are the following ones:

AXIOM OF EXTENSIONALITY:
Two sets X and Y coincide if and only if they have the same elements.

AXIOM OF SPECIFICATION (or AXIOM OF SEPARATION):
For each set X and property ϕ (expressed in the language of set theory) there exists a set Y whose elements are exactly the members of X that satisfy ϕ.

AXIOM OF CHOICE:
For every family X of non-null sets, one can *choose* an element from each set in X, i.e., there exists a function f on X such that $f(s)$ is a member of s, for every s in X.

---

[1]Indeed, if x is any member of $X = \{x_0, x_1, \ldots\}$, then $x = x_i$ for some natural number i. But then $x_{i+1} \in x \cap X$.

© Springer International Publishing AG, part of Springer Nature 2018
D. Cantone and P. Ursino, *An Introduction to the Technique of Formative Processes in Set Theory*, https://doi.org/10.1007/978-3-319-74778-1_1

The definition of the von Neumann standard cumulative hierarchy is based on the principle of transfinite recursion on ordinals, which is a generalization of the recursion principle on integers. Thus, in our introduction to set theory we shall also review the basic notions on ordinal numbers (for a more comprehensive presentation, see [Jec78]).

To begin with, we denote:

- by $f \colon X \to Y$, a function from X into Y;

- by Y^X, the set of all functions (or maps) f from X into Y;

- by $f[S]$, the multi-image $\{w \mid (\exists z \in S)w = f(z)\}$;

- by $\mathsf{dom}(f)$, the *domain* of f, namely, the collection of all x's for which $f(x)$ is defined (thus, if $f \in Y^X$, then $\mathsf{dom}(f) = X$);

- by $\mathsf{ran}(f)$, the *range* of f, namely, the collection of all y's for which $f(x) = y$, for some $x \in \mathsf{dom}(f)$ (thus, if $f \in Y^X$, then $\mathsf{ran}(f) \subseteq Y$);[2]

- by ι_X, the *identity function* on X, i.e., $\iota_X := \{\langle x, x \rangle \mid x \in X\}$ (where $\langle s, t \rangle$ denotes the ordered pair[3] of s and t); and

- by $f \circ g$, the *composition*[4] of f and g, where $f \colon X \to Y$ and $g \colon Y \to Z$.

To describe a function $f = \{\langle x, y_x \rangle \mid x \in X\}$, we shall interchangeably use the notations $f = \{y_x\}_{x \in X}$ and $x \overset{f}{\mapsto} y_x$ $(x \in X)$. When the function f is understood, we shall also just write $x \mapsto y_x$. A function $f \colon X \to Y$ is said to be

- *partial*, if f is a function from a possibly proper subset X' of X into Y; if f is partial, we shall also write either $f \colon X \nrightarrow Y$, or $x \overset{f}{\nmapsto} y_x$, or simply $x \mapsto y_x$ (when the function f is understood).

- *injective*, if $f(x') = f(x'')$ implies $x' = x''$, for all $x', x'' \in X$; if f is injective, we shall also write $f \colon X \rightarrowtail Y$;

- *surjective*, if $\mathsf{ran}(f) = Y$;

- *bijective*, if it is injective and surjective.

Injection, surjection, and bijection are synonyms of injective, surjective, and bijective function, respectively.

We agree on the following notations for the POWERSET, UNION SET, DISJOINT UNION SET, and INTERSECTION SET of a given a set T:

$$\mathcal{P}(T) := \{S \mid S \subseteq T\} \qquad \text{(the } powerset \text{ of } T)$$

$$\bigcup T := \{s \mid s \in t \text{ for some } t \in T\} \quad \text{(the } union\ set \text{ of } T)$$

$$\biguplus T := \{s \mid (\exists! t)(t \in T \wedge s \in t)\} \quad \text{(the } disjoint\ union\ set \text{ of } T)^5$$

$$\bigcap T := \{s \mid s \in t \text{ for every } t \in T\} \quad \text{(the } intersection\ set \text{ of } T, \text{ provided that } T \neq \emptyset).$$

---

[2] Notice that, for a function f, we have: $\mathsf{ran}(f) = f[\mathsf{dom}(f)]$.

[3] Order pairs can be encoded by sets. The standard encoding is due to the logician K. Kuratowski: $\langle s, t \rangle := \{\{s\}, \{s, t\}\}$.

[4] Throughout the book, we shall use functional-order notation. Thus, $f \circ g := \{\langle x, z \rangle \mid g(x) = y \text{ and } f(y) = z, \text{ for some } y\}$.

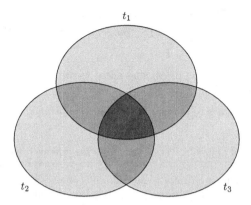

Figure 1.1: The sets $\bigcup T$, $\biguplus T$, and $\bigcap T$ relative to a three-element set $T = \{t_1, t_2, t_3\}$: $\bigcup T$ is the union of all the grey regions (of any darkness), $\biguplus T$ is the union of the light grey regions only, and, finally, $\bigcap T$ is the central dark grey region.

Figure 1.1 depicts the sets $\bigcup T$, $\biguplus T$, and $\bigcap T$ for a three-element set $T = \{t_1, t_2, t_3\}$. Specifically, $\bigcup T$ is the union of all the grey regions (of any darkness), $\biguplus T$ is the union of the light grey regions only, and, finally, $\bigcap T$ is the central dark grey region.

Remark 1.1. We have:

- $\biguplus\{t_1\} = t_1$;

- $\biguplus\{t_1, t_2\} = (t_1 \cup t_2) \setminus (t_1 \cap t_2)$;

- $\biguplus\{t_1, t_2, t_3\} = (t_1 \cup t_2 \cup t_3) \setminus \bigcup_{1 \leqslant i < j \leqslant 3}(t_i \cap t_j)$;

and, more in general,

- $\biguplus\{t_1, \ldots, t_n\} = (t_1 \cup \ldots \cup t_n) \setminus \bigcup_{1 \leqslant i < j \leqslant n}(t_i \cap t_j)$, for every $n \in \omega$, and

- $\biguplus T = \bigcup T \setminus \bigcup\{t \cap t' \mid t, t' \in T, \ t \neq t'\}$, for every set T. ∎

The following lemma collects the main properties of the operators $\mathcal{P}$, $\bigcup$, and $\bigcap$ relevant to our later developments:

Lemma 1.2. *Let S, T be sets such that $S \subseteq T$. Then the following properties hold:*

(a) $\mathcal{P}(S) \subseteq \mathcal{P}(T)$;

(b) $\bigcup S \subseteq \bigcup T$;

(c) *if $S \neq \emptyset$, then $\bigcap T \subseteq \bigcap S$;*

---

[5]By "∃!" we mean the *unique existential quantification*; thus, $(\exists! t)(t \in T \wedge s \in t)$ has to be interpreted as "there exists a unique t such that $t \in T \wedge s \in t$".

(d) $\bigcup \mathcal{P}(S) = S \subseteq \mathcal{P}(\bigcup S)$;

(e) $\mathcal{P}(\bigcup S) = S$ if and only if $S = \mathcal{P}(\bigcup S')$ for some set S';

(f) if $T = \mathcal{P}(\bigcup T)$, then $\mathcal{P}(\bigcup S) \subseteq T$;

(g) if $S \neq \emptyset$, then $\mathcal{P}(\bigcap S) = \bigcap \mathcal{P}[S]$, where $\mathcal{P}[\bullet]$ is the multi-powerset operator, i.e., $\mathcal{P}[S] = \{\mathcal{P}(s) \mid s \in S\}$;

(h) $\bigcup \bigcup S = \bigcup \bigcup [S]$, where $\bigcup [\bullet]$ is the multi-unionset operator, i.e., $\bigcup [S] = \{\bigcup s \mid s \in S\}$. ■

Properties (a) and (b) of Lemma 1.2 state that both the powerset operator and the union set operator are monotonically nondecreasing (with respect to set inclusion), whereas (c) states that the intersection set operator is monotonically nonincreasing. From (d), it follows that the union set operator is a left-inverse of the powerset operator. Instead, the powerset operator is a left-inverse of the union set operator only for sets of the form $\mathcal{P}(\bigcup S')$, for some S' (cf. property (e)). Properties (d), (e), and (f) entail that $\mathcal{P}(\bigcup S)$ is the minimal superset R of S such that $R = \mathcal{P}(\bigcup R)$ holds. Property (g) yields that, in some sense, the powerset and the intersection set operators commute. Finally, (h) implies that the union set operator distributes over the binary union operator. Indeed, for $S := \{S_1 \ldots S_n\}$, property (h) yields

$$\bigcup (S_1 \cup \ldots \cup S_n) = \bigcup \bigcup \{S_1, \ldots, S_n\}$$
$$= \bigcup \bigcup [\{S_1, \ldots, S_n\}] = \bigcup \{\bigcup S_1, \ldots, \bigcup S_n\} = \bigcup S_1 \cup \ldots \cup \bigcup S_n.$$

Remark 1.3. Note that $X \neq \mathcal{P}(X)$. Indeed, one may simply observe that $X \in \mathcal{P}(X)$, whereas $X \notin X$ by the well-foundedness of $\in$. The same conclusion can be reached without recourse to the Axiom of Regularity, anyhow, through the Axiom of Specification (cf. [Zer77]). The latter enables us to build the set $S := \{v \in X \mid v \notin v\}$, which plainly fulfills $S \in \mathcal{P}(X)$, and $(\forall v \in S)(v \notin v)$—and hence $S \notin S$. Let $T := S \cup \{S\}$. Then we have:

(i) $S \in \mathcal{P}(T)$, and

(ii) $(\forall v \in T)(v \notin v)$.

In addition, since $S \in T$ and $S \notin S$, then $S \neq T$, and therefore the Axiom of Extensionality entails

(iii) $S \notin \mathcal{P}(T) \vee T \notin \mathcal{P}(S)$.

Assume now by contradition that $X = \mathcal{P}(X)$ holds. Since $S \in \mathcal{P}(X)$, then $S \in X$, so that $T \in \mathcal{P}(X)$. Thus, by (ii), $T \in \mathcal{P}(S)$. In view of (i), the latter contradicts (iii), proving that $X \neq \mathcal{P}(X)$. ■

Definition 1.4. *A* LINEAR ORDER *on a set X is any binary relation on X which is transitive, antisymmetric, and total.*[6] *In addition, a linear order $\leqslant$ on X such that every non-null subset Y of X has a least element with respect to $\leqslant$ is a* WELL-ORDER. ■

---

[6]We recall that a binary relation R on X is: *reflexive*, if $x\,\mathsf{R}\,x$ holds for every $x \in X$; *transitive*, if $x\,\mathsf{R}\,y$ and $y\,\mathsf{R}\,z$ yield $x\,\mathsf{R}\,z$; *symmetric*, if $x\,\mathsf{R}\,y$ yields $y\,\mathsf{R}\,x$; *antisymmetric*, if $x\,\mathsf{R}\,y$ and $y\,\mathsf{R}\,x$ yield $x = y$; *total*, if $x\,\mathsf{R}\,y$, or $x = y$, or $y\,\mathsf{R}\,x$, for all $x, y \in X$. We also recall that a binary relation is an EQUIVALENCE RELATION if it is reflexive, symmetric, and transitive.

Transitive sets and ordinal numbers

We say that a set s PARTICIPATES IN THE CONSTRUCTION of another set T if there are sets $s_0, s_1, \ldots, s_n$, with $n \geqslant 0$, such that

$$s = s_0 \in s_1 \in \cdots \in s_n \in T.$$

A set which contains as members all the elements which participate in its construction is said to be TRANSITIVE. More formally, transitivity can be defined as follows:

Definition 1.5. *A set T is* TRANSITIVE *if, whenever $r \in s$ and $s \in T$, then $r \in T$ holds.*

Some characterizations of transitive sets are listed in the following lemma.

Lemma 1.6. *For a set T, the following conditions are equivalent:*

(a) T is transitive;

(b) $T \subseteq \mathcal{P}(T)$;

(c) $\bigcup T \subseteq T$.

Proof. We show that (a) $\Rightarrow$ (b) $\Rightarrow$ (c) $\Rightarrow$ (a). Let us assume first that T is transitive and let $s \in T$. Then for each $r \in s$, by the transitivity of T, we have $r \in T$, so that $s \subseteq T$, i.e., $s \in \mathcal{P}(T)$, proving (a) $\Rightarrow$ (b). Next, assume that $T \subseteq \mathcal{P}(T)$ holds. By applying the union set operator to both sides of the inclusion, we obtain $\bigcup T \subseteq \bigcup \mathcal{P}(T) = T$, proving (b) $\Rightarrow$ (c). Finally, assume that $\bigcup T \subseteq T$ holds, and let $r \in s \in T$. Then $r \in \bigcup T$, so that, by our assumption, $r \in T$, proving (c) $\Rightarrow$ (a), and in turn proving that conditions (a), (b), and (c) of the lemma are equivalent. $\qquad\square$

The following lemma collects some elementary properties of transitive sets.

Lemma 1.7. *For a set S and a family $\mathcal{T}$ of transitive sets, the following properties hold:*

(a) $\bigcup \mathcal{T}$ is transitive;

(b) if $\mathcal{T} \neq \emptyset$, then $\bigcap \mathcal{T}$ is transitive;

(c) if S is transitive, then so is $\bigcup S$;

(d) S is transitive if and only if so is $\mathcal{P}(S)$;

(e) if S is a non-null transitive set, then $\emptyset \in S$.

Proof. Let S be a set and $\mathcal{T}$ a family of transitive sets.

For every $T \in \mathcal{T}$, we have $T \subseteq \bigcup \mathcal{T}$. Hence, by the transitivity of T and the monotonicity of the powerset operator, we have $T \subseteq \mathcal{P}(T) \subseteq \mathcal{P}(\bigcup \mathcal{T})$, so that $\bigcup \mathcal{T} = \bigcup_{T \in \mathcal{T}} T \subseteq \mathcal{P}(\bigcup \mathcal{T})$, proving the transitivity of $\bigcup \mathcal{T}$; therefore (a) holds.

As for (b), let now $\mathcal{T} \neq \emptyset$. Then, for every $T \in \mathcal{T}$, we have $\bigcap \mathcal{T} \subseteq T \subseteq \mathcal{P}(T)$, so that, by Lemma 1.2(g),

$$\bigcap \mathcal{T} \subseteq \bigcap_{T \in \mathcal{T}} \mathcal{P}(T) = \bigcap \mathcal{P}[\mathcal{T}] = \mathcal{P}(\bigcap \mathcal{T}),$$

proving that $\bigcap \mathcal{T}$ is transitive.

Concerning (c) and (d), if S is transitive then, by Lemma 1.6, $\bigcup S \subseteq S$ and $S \subseteq \mathcal{P}(S)$. By applying the union set operator to both sides of the former, we obtain $\bigcup\bigcup S \subseteq \bigcup S$, proving that $\bigcup S$ is transitive, so that (c) holds. From the latter and Lemma 1.2, it follows $\bigcup \mathcal{P}(S) = S \subseteq \mathcal{P}(S)$, proving that $\mathcal{P}(S)$ is transitive, whenever S is transitive. Conversely, if $\mathcal{P}(S)$ is transitive, then, by Lemma 1.2(d), we have $S = \bigcup \mathcal{P}(S)$, so that, by property (c) just proved, S is transitive, completing the proof of (d).

Finally, if S is a non-null transitive set, by the Axiom of Regularity, S has a member s not intersecting S. Hence, by transitivity, we have $s = s \cap S = \emptyset$, proving (e). □

Definition 1.8. *A set μ is said to be an* ORDINAL NUMBER *if it is transitive and is linearly ordered by the membership relation $\in$ (and hence well-ordered, by the Axiom of Regularity).* ∎

As is customary, ordinal numbers will be denoted by small Greek letters α, β, γ, ...

Example 1.9. The following sets are ordinal numbers (and therefore transitive as well):

$$\emptyset, \quad \{\emptyset\}, \quad \{\emptyset, \{\emptyset\}\}, \quad \{\emptyset, \{\emptyset\}, \{\emptyset, \{\emptyset\}\}\}.$$

Instead, the set $\{\emptyset, \{\emptyset\}, \{\{\emptyset\}\}, \{\emptyset, \{\emptyset\}\}\}$ is transitive, but not an ordinal number, since neither $\{\{\emptyset\}\} \in \{\emptyset, \{\emptyset\}\}$, nor $\{\emptyset, \{\emptyset\}\} \in \{\{\emptyset\}\}$ holds (that is, completeness fails).

Finally, the set $\{\emptyset, \{\emptyset, \{\emptyset\}\}\}$ is not transitive (hence, it fails to be an ordinal number) since though $\{\emptyset\} \in \{\emptyset, \{\emptyset\}\} \in \{\emptyset, \{\emptyset, \{\emptyset\}\}\}$, we have $\{\emptyset\} \notin \{\emptyset, \{\emptyset, \{\emptyset\}\}\}$, where, plainly, the set $\{\emptyset\}$ participates in the construction of $\{\emptyset, \{\emptyset, \{\emptyset\}\}\}$. ∎

As the membership relation $\in$ is a well-ordering of the class On of all ordinals, for every non-null set $C \subseteq On$, we have $\bigcap C \in C$, and therefore $\bigcap C$ is the smallest ordinal in C, i.e., $\bigcap C = \min C$. Also, for every non-null set $C \subseteq On$, we have $\bigcup C \in On$, and therefore $\bigcup C$ is the smallest ordinal γ such that, for every $\alpha \in C$, either $\alpha = \gamma$ or $\alpha \in \gamma$, i.e., $\bigcup C = \sup C$.

One reason to be interested in ordinals is the following fundamental theorem:

Theorem 1.10. *Let $\trianglelefteq$ be a well-ordering on the set x. Then there exist, and are uniquely determined, an ordinal ξ and a function $f \in x^{\xi}$ such that $f[\xi] = x$ holds and, for any pair $\nu, \mu < \xi$ of ordinals:*

$$f\nu \neq f\mu \text{ holds when } \nu \neq \mu, \text{ and moreover } f\nu \trianglelefteq f\mu \text{ when } \nu \leq \mu.$$

By the Axiom of Choice, a well-ordering can be imposed on any set (cf. [Jec78]). Therefore the following definition makes sense: the CARDINALITY of a set S, denoted by $|S|$, is the least ordinal ν such that there exists a function $f \colon \nu \to S$ for which $f[\nu] = S$ holds. A CARDINAL NUMBER is an ordinal μ such that $\mu = |\mu|$.

It is rather easy to see in which sense ordinals are an extension of the natural numbers. Indeed, natural numbers, defined *à la* von Neumann by the rules

$$0 := \emptyset, \quad i + 1 := i \cup \{i\}, \tag{1.2}$$

constitute the initial segment of the class of ordinals: thus, $n = \{0, 1, \ldots, n-1\}$ for each natural number n. Their collection, $\omega := \{0, 1, 2, 3, \ldots\}$, is the first ordinal which exceeds all natural numbers, often denoted $\aleph_0$. The set of all finite ordinals will also be denoted by $\mathbb{N}$.

Even for ordinals (such as ω) which are not natural numbers, it is convenient to assign the meaning just indicated to the increment operation '+1': we shall hence have, among ordinals, $\omega + 1$, $\omega + 1 + 1$, etc. The ordinals of the form $\mu + 1$, where μ is an ordinal, are called SUCCESSOR ORDINALS; all others, save 0, are INFINITE LIMIT ORDINALS. The latter comprise ω, $\omega + \omega$, $\underbrace{\omega + \omega + \ldots}_{\omega \text{ times}}$, etc. (we are appealing to the intuition of the reader). Finally, we agree that the ordinal 0 is the unique FINITE LIMIT ORDINAL.

All elements of $\omega + 1$ are cardinal numbers; but $\omega + 1$ itself is *not* a cardinal number.

Definition 1.11. *By a ξ-SEQUENCE, where ξ is an ordinal, we mean a function $\mu \mapsto Y_\mu$ ($\mu \in \xi$), usually denoted $(Y_\mu)_{\mu < \xi}$, whose domain is ξ. When the ordinal ξ is omitted, by a SEQUENCE we mean either an ω-sequence or an n-sequence, with $n \in \omega$, depending on context.* ∎

The von Neumann standard cumulative hierarchy

The von Neumann standard cumulative hierarchy $\boldsymbol{V}$ of all sets is defined by the following transfinite recursion on the class On of all ordinals:

$$\begin{aligned}
\boldsymbol{V}_0 &= \emptyset \\
\boldsymbol{V}_{\alpha+1} &= \mathcal{P}(\boldsymbol{V}_\alpha), && \text{for each ordinal } \alpha, \\
\boldsymbol{V}_\lambda &= \bigcup_{\mu < \lambda} \boldsymbol{V}_\mu, && \text{for each limit ordinal } \lambda, \\
\boldsymbol{V} &= \bigcup_{\mu \in On} \boldsymbol{V}_\mu.
\end{aligned}$$

By transfinite recursion and Lemma 1.7, it can be shown that each layer $\boldsymbol{V}_\alpha$ of the von Neumann cumulative hierarchy is transitive, for every ordinal α.

Example 1.12. Using the following 'nesting' notation

$$\emptyset^n := \underbrace{\{\{\ldots\{}_{n} \emptyset \underbrace{\}\ldots\}\}}_{n},$$

the first few layers of the von Neumann standard cumulative hierarchy are

$$
\begin{aligned}
\mathcal{V}_0 &= \emptyset \\
\mathcal{V}_1 &= \{\emptyset\} \\
\mathcal{V}_2 &= \{\emptyset, \emptyset^1\} \\
\mathcal{V}_3 &= \{\emptyset, \emptyset^1, \emptyset^2, \{\emptyset, \emptyset^1\}\} \\
\mathcal{V}_4 &= \big\{\emptyset, \\
&\quad \emptyset^1, \emptyset^2, \emptyset^3, \{\{\emptyset, \emptyset^1\}\}, \\
&\quad \{\emptyset, \emptyset^1\}, \{\emptyset, \emptyset^2\}, \{\emptyset, \{\emptyset, \emptyset^1\}\}, \{\emptyset^1, \emptyset^2\}, \{\emptyset^1, \{\emptyset, \emptyset^1\}\}, \{\emptyset^2, \{\emptyset, \emptyset^1\}\}, \\
&\quad \{\emptyset, \emptyset^1, \emptyset^2\}, \{\emptyset, \emptyset^1, \{\emptyset, \emptyset^1\}\}, \{\emptyset, \emptyset^2, \{\emptyset, \emptyset^1\}\}, \{\emptyset^1, \emptyset^2, \{\emptyset, \emptyset^1\}\}, \\
&\quad \{\emptyset, \emptyset^1, \emptyset^2, \{\emptyset, \emptyset^1\}\}\big\}
\end{aligned}
$$

$$\vdots \ \vdots \qquad\qquad \vdots$$

(Notice that the elements in $\mathcal{V}_4$ have been grouped by cardinalities.) ■

For any given set S, the maximal depth of nesting of the empty set $\emptyset$ in S is an important measure related to S, called *rank*.

Definition 1.13. *The* RANK *of a set* $S \in \mathcal{V}$, *denoted by* rk S, *is the least ordinal* α *such that* $S \in \mathcal{V}_{\alpha+1}$. ■

The class $\mathcal{V}_\mu$ of all sets whose rank is smaller than μ is, for every ordinal μ, a *set*, which is easily recognized to be transitive. Among these sets, one has the family $\mathcal{V}_\omega$ (also denoted by HF) of the HEREDITARILY FINITE sets, which are those sets that are *finite* and whose elements, elements of elements, etc., all are finite.

Example 1.14. The sets $\{\{\emptyset, \emptyset^1, \emptyset^2, \ldots\}\}$ and $\{\mathcal{V}_\omega\}$ are simple examples of finite, but not hereditarily finite sets. Indeed, their cardinality is 1, and both of them contain an infinite element. ■

The following lemma lists some useful properties of the rank function (for a proof, see [CFO89]).

Lemma 1.15. *For all sets* $S, T, S_0, \ldots, S_n \in \mathcal{V}$ *and every ordinal* α, *we have:*

  (i) rk$(S \cup T) = \max\{\text{rk } S, \text{rk } T\}$;

  (ii) rk $\emptyset = 0$ *and* rk $\{S_0, \ldots, S_n\} = \max\{\text{rk } S_0, \ldots, \text{rk } S_n\} + 1$;

  (iii) rk $S = \bigcup\{\text{rk } U + 1 \mid U \in S\}$;

  (iv) *if* $S \in T$, *then* rk $S < $ rk T;

  (v) *if* $S \subseteq T$, *then* rk $S \leqslant$ rk T;

  (vi) rk $\bigcup S = \bigcup\{\text{rk } U \mid U \in S\}$;

  (vii) $\mathcal{V}_\alpha = \{S \in \mathcal{V} \mid \text{rk } S \in \alpha\}$;

(viii) $\text{rk } \mathbf{V}_\alpha = \text{rk } \alpha = \alpha$;

(ix) there exists a well-ordering $\trianglelefteq$ on S such that

$$T \trianglelefteq U \quad when \quad \text{rk } T < \text{rk } U, \quad for \quad T, U \in S.$$

Hereditarily finite sets can be conveniently represented as finite downward growing trees, as in the next picture:

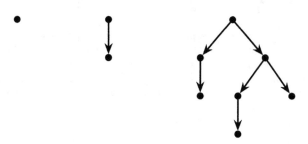

By traversing trees in a bottom-up fashion, one can compute the sets they represent through the following labeling process:

(i) all leaves are initially labeled with the empty set $\emptyset$; then

(ii) each internal node is recursively labeled with the collection of the labels of its children; finally,

(iii) the root label is the set represented by the tree.

The following picture illustrates the labeling process just described on three trees:[7]

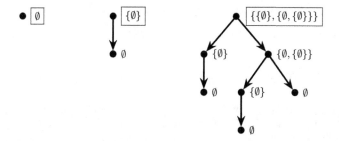

However, the same set may be represented by different trees, as in the picture:

---

[7]The labeling process works also for infinite *well-founded trees*, namely, trees whose branches have finite length.

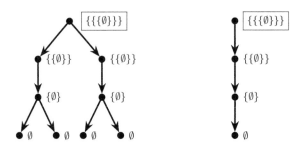

To circumvent such a problem, it is possible to define an equivalence relation $\approx$ among trees, called MAXIMUM BISIMULATION, such that any set is represented by a unique $\approx$-equivalence class in the collection of all trees; the interested reader can find a thorough treatment of this subject in [Acz88].

As can be easily proved by induction, the rank of a hereditarily finite set coincides with the *height* of any tree that represents it:[8]

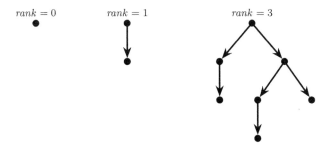

Thus, $\mathsf{rk}\,\emptyset = 0$, $\mathsf{rk}(\{\emptyset\}) = 1$, and $\mathsf{rk}(\{\{\emptyset\}, \{\emptyset, \{\emptyset\}\}\}) = 3$, etc.

Transitive closure

Any set S can be uniquely associated with a transitive set, as in the following definition.

Definition 1.16. *The* TRANSITIVE CLOSURE $\mathsf{TrCl}(S)$ *of a set S is the minimal transitive set containing S, i.e.,*

$$\mathsf{TrCl}(S) := \bigcap \{T \in \boldsymbol{\mathcal{V}} \mid S \subseteq T \wedge T \text{ is transitive}\}. \qquad \blacksquare$$

The definition of $\mathsf{TrCl}(S)$ is well-given. Indeed, if S is a set, then $S \in \boldsymbol{\mathcal{V}}_{\overline{\alpha}}$, for some ordinal $\overline{\alpha}$. Hence $\{T \in \boldsymbol{\mathcal{V}} \mid S \subseteq T \wedge T \text{ is transitive}\}$ is non-null, as it plainly contains $\boldsymbol{\mathcal{V}}_{\overline{\alpha}}$.

The following lemma lists some elementary properties connected to transitive closure:

Lemma 1.17. *The following properties hold for any set S:*

(a) $\mathsf{TrCl}(S)$ *is transitive;*

---

[8] We recall that the *height* of a tree is the maximum length of a path in the tree from the root to a leaf.

(b) S is transitive if and only if $\mathsf{TrCl}(S) = S$;

(c) a set is hereditarily finite if and only if its transitive closure is finite.

Partitions, transitive partitions, and $\mathcal{P}$-partitions

The notion of *partition*, and its variants, will be basic to our later developments.

Definition 1.18. *A* PARTITION Σ *is a collection of pairwise disjoint non-null sets, called the* BLOCKS *of Σ. The union $\bigcup\Sigma$ of a partition Σ is the* DOMAIN *of Σ. A partition is* TRANSITIVE *if so is its domain. Likewise, a partition is* HEREDITARILY FINITE *if so is its domain.* ∎

Notice that the domain of a transitive partition contains all the elements that participate in its construction (cf. Exercise 1.3).

If a partition Σ is non-transitive, then we can obtain a transitive partition containing Σ as a subset, by adding to it as a new block the set $\mathsf{TrCl}(\bigcup\Sigma) \setminus \bigcup\Sigma$ (which is plainly non-null, as $\bigcup\Sigma$ is a non-transitive subset of the transitive set $\mathsf{TrCl}(\bigcup\Sigma)$). In fact, it can be shown that $\Sigma \cup \{\mathsf{TrCl}(\bigcup\Sigma) \setminus \bigcup\Sigma\}$ is the minimal transitive partition containing Σ as a subset.

Definition 1.19. *The* TRANSITIVE COMPLETION *of a non-transitive partition Σ is the partition $\Sigma \cup \{\mathsf{TrCl}(\bigcup\Sigma) \setminus \bigcup\Sigma\}$.*

We agree that the transitive completion of a transitive partition Σ is the partition itself. ∎

For technical reasons, it will be more convenient to consider a stronger kind of completion. Let Σ be a partition, and σ^* any (non-null) set disjoint from $\bigcup\Sigma$ and such that the following chain of inclusions and equalities are fulfilled:

$$\bigcup\Sigma \subseteq \bigcup\bigcup(\Sigma \cup \{\sigma^*\}) \subseteq \mathcal{P}\left(\bigcup\bigcup(\Sigma \cup \{\sigma^*\})\right) = \bigcup(\Sigma \cup \{\sigma^*\}). \qquad (1.3)$$

Then, setting $\Sigma^* := \Sigma \cup \{\sigma^*\}$, we have:

(I) Σ^* is a transitive partition, and

(II) $\mathcal{P}(\bigcup\Gamma) \subseteq \bigcup\Sigma^*$, for each $\Gamma \subseteq \Sigma$.

Upon calling INTERNAL the blocks in Σ^* distinct from σ^*, constraint (II) just asserts that the domain of the partition Σ^* contains the powerset of every possible union of internal blocks. This closure property, together with a similar property involving the *intersecting powerset operator* $\mathcal{P}^*$ to be defined later, will be particularly useful in the applications of certain structures, called syllogistic boards,[9] to the decision problem in set theory.

Before proceeding any further, we need to check that, for any partition Σ and any set σ^* satisfying (1.3), constraints (I) and (II) are satisfied. In addition, we have to show that there actually exists some non-null set σ^* disjoint from $\bigcup\Sigma$ and fulfilling (1.3). We prove these facts in the following two lemmas.

---

[9]Syllogistic boards are graph structures superimposed to partitions to support their construction step by step by means of formative processes. Syllogistic boards and formative processes will be defined and elaborated in Chapter 3.

Lemma 1.20. *Let Σ be any partition, and let σ^* be any non-null set disjoint from $\bigcup \Sigma$ and such that (1.3) is fulfilled. Then, constraints (I) and (II) above are satisfied.*

Proof. Let Σ and σ^* be as in the statement. Then, $\bigcup\bigcup \Sigma^* \subseteq \bigcup \Sigma^*$ follows at once, so that, by Lemma 1.6(c), the domain $\bigcup \Sigma^*$ (and, therefore, the partition Σ^*) is transitive, proving that the constraint (I) is fulfilled. Concerning the second constraint, let $\Gamma \subseteq \Sigma$. Then, (1.3) yields

$$\bigcup \Gamma \subseteq \bigcup \Sigma \subseteq \bigcup\bigcup \Sigma^*,$$

so that, again by (1.3), we have

$$\mathcal{P}\left(\bigcup \Gamma\right) \subseteq \mathcal{P}\left(\bigcup\bigcup \Sigma^*\right) = \bigcup \Sigma^*,$$

proving (II). $\square$

Lemma 1.21. *Let Σ be any partition, and let*

$$\sigma^* := \mathcal{P}(\mathsf{TrCl}(\textstyle\bigcup \Sigma)) \setminus \bigcup \Sigma. \tag{1.4}$$

Then, σ^* is non-null and fulfills (1.3).

Proof. Given a partition Σ, let σ^* be defined as in (1.4). Then σ^* is plainly disjoint from $\bigcup \Sigma$, and is also non-null, as $\mathsf{TrCl}(\bigcup \Sigma) \in \sigma^*$. In addition, we have

$$\bigcup\bigcup(\Sigma \cup \{\sigma^*\}) = \bigcup \left(\mathcal{P}(\mathsf{TrCl}(\textstyle\bigcup \Sigma))\right) = \mathsf{TrCl}(\textstyle\bigcup \Sigma),$$

so that, by (1.4) and the transitivity of $\bigcup\bigcup(\Sigma \cup \{\sigma^*\})$, the inclusions and equality in (1.3) follow. $\square$

We shall interchangeably refer to the block σ^* (sometimes denoted also by $\bar{\sigma}$) either as the EXTERNAL BLOCK OF Σ (relative to the $\mathcal{P}$-completion Σ^*) or as the SPECIAL BLOCK OF Σ^*. Additionally, we shall refer to the partition Σ^* as the $\mathcal{P}$-COMPLETION OF Σ.

Definition 1.22 ($\mathcal{P}$-partition). *A partition Σ is said to be a $\mathcal{P}$-PARTITION if it contains a block σ^* such that*

$$\bigcup(\Sigma \setminus \{\sigma^*\}) \subseteq \bigcup\bigcup \Sigma \subseteq \mathcal{P}\left(\bigcup\bigcup \Sigma\right) = \bigcup \Sigma. \tag{1.5}$$

As is expected, the $\mathcal{P}$-completion of a given partition Σ is a $\mathcal{P}$-partition. The proof of this simple fact is left to the reader as an exercise.

It is not hard to show that if Σ is a $\mathcal{P}$-partition, then there is *exactly* one block $\sigma^* \in \Sigma$ fulfilling conditions (1.5) (the reader is asked to prove this fact in Exercise 1.12). We shall refer to this block as the SPECIAL BLOCK OF Σ. This will generate no confusion, since, as can be easily checked, if σ^* is the special block of the $\mathcal{P}$-completion Σ^* of a given partition Σ (in the sense that σ^* is the external block of Σ), then σ^* is also the special block of Σ^*, considered as a $\mathcal{P}$-partition.

Some elementary, yet useful properties of partitions and $\mathcal{P}$-partitions are stated in the following lemmas.

Lemma 1.23. *Let Σ be a partition. Then, for every $\sigma \in \Sigma$ and $\Gamma \subseteq \Sigma$, we have*

(a) $\sigma \in \Gamma$ if and only if $\sigma \cap \bigcup \Gamma \neq \emptyset$;

(b) $\sigma \in \Gamma$ if and only if $\sigma \subseteq \bigcup \Gamma$.

Proof. Let $\sigma \in \Sigma$ and $\Gamma \subseteq \Sigma$. Plainly, if $\sigma \in \Gamma$, then $\sigma \subseteq \bigcup \Gamma$, so that $\sigma \cap \bigcup \Gamma \neq \emptyset$. Conversely, if $\sigma \cap \bigcup \Gamma \neq \emptyset$, let $\sigma' \in \Gamma$ be such that $\sigma \cap \sigma' \neq \emptyset$. Since Σ is a partition, it follows that $\sigma = \sigma'$, so that $\sigma \in \Gamma$, proving (a).

Property (b) can be proved similarly to (a). $\qquad\square$

Lemma 1.24. *Let Σ be a $\mathcal{P}$-partition with special block σ^*. Then:*

(a) *Σ is transitive;*

(b) *$\mathcal{P}(\bigcup \Gamma) \subseteq \bigcup \Sigma$, for every $\Gamma \subseteq \Sigma \setminus \{\sigma^*\}$;*

(c) *$\bigcup(\Sigma \setminus \{\sigma\}) \in \sigma$ if and only if $\sigma = \sigma^*$, for every $\sigma \in \Sigma$.*

Proof. We prove here only property (c) and leave the proof of the remaining properties to the reader as an exercise. Thus, as in the hypothesis of the lemma, let Σ be a $\mathcal{P}$-partition with special block σ^*. Since $\bigcup(\Sigma \setminus \{\sigma^*\}) \in \bigcup \Sigma$, then $\bigcup(\Sigma \setminus \{\sigma^*\}) \in \sigma$, for some $\sigma \in \Sigma$. However, if $\sigma \neq \sigma^*$, then $\sigma \subseteq \bigcup(\Sigma \setminus \{\sigma^*\})$, so that, by properties (iv) and (v) Lemma 1.15, we would have

$$\operatorname{rk} \sigma \leqslant \operatorname{rk}(\bigcup(\Sigma \setminus \{\sigma^*\})) < \operatorname{rk} \sigma,$$

which is a contradiction. Hence, $\bigcup(\Sigma \setminus \{\sigma^*\}) \in \sigma^*$.

Conversely, let $\bigcup(\Sigma \setminus \{\sigma\}) \in \sigma$, for some $\sigma \in \Sigma$. If $\sigma \neq \sigma^*$, then

$$\sigma \subseteq \bigcup(\Sigma \setminus \{\sigma^*\}) \in \sigma^* \subseteq \bigcup(\Sigma \setminus \{\sigma\}) \in \sigma$$

and, once more by (iv) and (v) of Lemma 1.15, we would get again the contradiction $\operatorname{rk} \sigma < \operatorname{rk} \sigma$. Therefore, $\sigma = \sigma^*$ must hold, completing the proof of property (c) of the lemma. $\qquad\square$

Venn partitions and some of its variants

In connection with the decision problem for fragments of set theory, and more specifically with the formative process technique, we shall be interested in the (transitive) Venn partitions and Venn $\mathcal{P}$-partitions of finite collections of sets. The next three definitions collect all relevant notions.

Let $\mathcal{F}$ be a finite collection of sets.

Definition 1.25. *The Venn partition of $\mathcal{F}$ is the partition $\Sigma_{\mathcal{F}}^{\mathsf{V}}$ induced by the equivalence relation $\sim_{\mathcal{F}}$ over $\bigcup \mathcal{F}$, where*

$$s \sim_{\mathcal{F}} t \quad \text{if and only if} \quad s \in S \leftrightarrow t \in S, \text{ for every } S \in \mathcal{F}. \qquad\blacksquare$$

As can be easily checked, we have

$$\Sigma_{\mathcal{F}}^{\mathsf{V}} = \left\{ \bigcap \mathcal{E} \setminus \bigcup(\mathcal{F} \setminus \mathcal{E}) \,\middle|\, \emptyset \neq \mathcal{E} \subseteq \mathcal{F} \right\} \setminus \{\emptyset\}.$$

Hence, $|\Sigma_{\mathcal{F}}^{\mathsf{V}}| \leqslant |\mathcal{P}(\mathcal{F}) \setminus \{\emptyset\}| = 2^{|\mathcal{F}|} - 1$. In addition, $\bigcup \Sigma_{\mathcal{F}}^{\mathsf{V}} = \bigcup \mathcal{F}$, so that $\Sigma_{\mathcal{F}}^{\mathsf{V}}$ is a transitive partition if and only if $\bigcup \mathcal{F}$ is transitive.

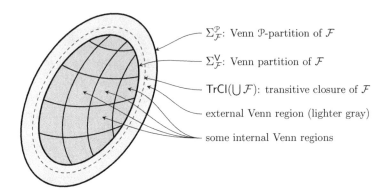

Figure 1.2: Venn $\mathcal{P}$-partition, Venn partition, and transitive closure of $\mathcal{F}$.

Definition 1.26. *The* TRANSITIVE VENN PARTITION *of* $\mathcal{F}$, *denoted by* $\Sigma_{\mathcal{F}}^{\mathsf{T}}$, *is the Venn partition of the extended collection* $\mathcal{F} \cup \{\mathsf{TrCl}(\bigcup \mathcal{F})\}$.[10]   ∎

It is easy to verify that we have

$$\Sigma_{\mathcal{F}}^{\mathsf{T}} = \begin{cases} \Sigma_{\mathcal{F}}^{\mathsf{V}} & \text{if } \Sigma_{\mathcal{F}}^{\mathsf{V}} \text{ is transitive} \\ \Sigma_{\mathcal{F}}^{\mathsf{V}} \cup \{\mathsf{TrCl}(\bigcup \mathcal{F}) \setminus \bigcup \mathcal{F}\} & \text{otherwise} \end{cases}$$

and $\bigcup \Sigma_{\mathcal{F}}^{\mathsf{T}} = \mathsf{TrCl}(\bigcup \mathcal{F})$, so that (the domain of) $\Sigma_{\mathcal{F}}^{\mathsf{T}}$ is transitive.[11] Hence, $\Sigma_{\mathcal{F}}^{\mathsf{T}}$ is a transitive partition which (possibly) extends $\Sigma_{\mathcal{F}}^{\mathsf{V}}$ and such that $\bigcup \Sigma_{\mathcal{F}}^{\mathsf{T}} = \mathsf{TrCl}(\bigcup \Sigma_{\mathcal{F}}^{\mathsf{V}})$ and $|\Sigma_{\mathcal{F}}^{\mathsf{T}}| \leqslant 2^{|\mathcal{F}|}$ hold.

Definition 1.27. *A* VENN $\mathcal{P}$-PARTITION $\Sigma_{\mathcal{F}}^{\mathcal{P}}$ *of* $\mathcal{F}$ *is any* $\mathcal{P}$-*completion of the Venn partition* $\Sigma_{\mathcal{F}}^{\mathsf{V}}$ *of* $\mathcal{F}$. *The blocks of* $\Sigma_{\mathcal{F}}^{\mathcal{P}}$ *belonging to* $\Sigma_{\mathcal{F}}^{\mathsf{V}}$ *are the* INTERNAL *or* PROPER VENN REGIONS *or* VENN PARTS *of* $\mathcal{F}$, *and the special block of* $\Sigma_{\mathcal{F}}^{\mathcal{P}}$ *is the* EXTERNAL *or* SPECIAL VENN REGION *of* $\mathcal{F}$ *(relative to the Venn* $\mathcal{P}$-*partition* $\Sigma_{\mathcal{F}}^{\mathcal{P}}$.)   ∎

(Figure 1.2 shows the Venn $\mathcal{P}$-partition, Venn partition, and transitive closure of $\mathcal{F}$, as well as their external and internal regions.) Plainly, $|\Sigma_{\mathcal{F}}^{\mathcal{P}}| \leqslant 2^{|\mathcal{F}|}$.

The Venn partition $\Sigma_{\mathcal{F}}^{\mathsf{V}}$ of a collection $\mathcal{F}$ enjoys the following elementary property:

for every $S \in \mathcal{F}$ there is a (unique) $\Gamma_S \subseteq \Sigma_{\mathcal{F}}^{\mathsf{V}}$ such that $S = \bigcup \Gamma_S$.

In fact, $\Gamma_S = \{\sigma \in \Sigma_{\mathcal{F}}^{\mathsf{V}} \mid \sigma \subseteq S\}$.

The intersecting powerset operator $\mathcal{P}^*$

The variant $\mathcal{P}^*$ of the powerset operator, introduced in [Can91] and recalled below, will also be particularly important in the rest of the book.

---

[10]We recall that TrCl is the transitive closure operator (see Definition 1.16).

[11]Notice that if $\bigcup \mathcal{F}$ is transitive, then $\Sigma_{\mathcal{F}}^{\mathsf{T}} = \Sigma_{\mathcal{F}}^{\mathsf{V}}$.

Definition 1.28. *For a given set S, the* INTERSECTING POWERSET *of S, denoted by $\mathcal{P}^*(S)$, is the collection of the subsets of $\bigcup S$ which have non-null intersection with* all *the members of S;[12] in symbols,*

$$\mathcal{P}^*(S) := \{ s \mid s \subseteq \textstyle\bigcup S \text{ and } s \cap t \neq \emptyset, \text{ for all } t \in S \}. \tag{1.6}$$

Thus, $\mathcal{P}^*(S) \subseteq \mathcal{P}(\bigcup S)$ and $\mathcal{P}^*(\{t\}) = \mathcal{P}(t) \setminus \{\emptyset\}$. In addition, $\mathcal{P}^*(S) = \emptyset$ whenever $\emptyset \in S$. Otherwise (namely, if $\emptyset \notin S$) $\bigcup S \in \mathcal{P}^*(S)$ and therefore $\bigcup \mathcal{P}^*(S) = \bigcup S$. In particular, $\mathcal{P}^*(\emptyset) = \{\emptyset\}$. We also have

$$\mathcal{P}(\textstyle\bigcup S) = \bigcup \{\mathcal{P}^*(T) \mid T \subseteq S\}. \tag{1.7}$$

A partition S induces in a natural way a partition of $\mathcal{P}(\bigcup S)$, as proved in the following lemma.

Lemma 1.29. *If S is a partition, then the family $\{\mathcal{P}^*(T) \mid T \subseteq S\}$ is a partition with domain $\mathcal{P}(\bigcup S)$.*

Proof. Let S be a partition. To prove that $\{\mathcal{P}^*(T) \mid T \subseteq S\}$ is a partition too, we just need to verify that

(i) $\emptyset \notin \{\mathcal{P}^*(T) \mid T \subseteq S\}$; and

(ii) $\mathcal{P}^*(T_1) \cap \mathcal{P}^*(T_2) = \emptyset$, for all $T_1, T_2 \subseteq S$ such that $T_1 \neq T_2$.

Property (i) is immediate, as S is a partition and therefore $\emptyset \notin S$. Concerning (ii), let $T_1, T_2 \subseteq S$ be such that $T_1 \neq T_2$ and, without loss of generality, let $t \in T_1 \setminus T_2$. But then, by the very definition (1.6) of the intersecting operator $\mathcal{P}^*$, we have that

- each element in $\mathcal{P}^*(T_1)$ intersects t;

- each element in $\mathcal{P}^*(T_2)$ is contained as a subset in $\bigcup T_2$ and, therefore, has empty intersection with t.

Hence (ii) follows.

Finally, by (1.7) above, it follows readily that the domain of $\{\mathcal{P}^*(T) \mid T \subseteq S\}$ is $\mathcal{P}(\bigcup S)$. $\qquad\square$

Example 1.30. Let us consider the partition $S := \{\{0,1\}, \{2,3\}\}$. Then we have

$$\mathcal{P}^*(\{\{0,1\}, \{2,3\}\}) = \{\{0,2\}, \{0,3\}, \{0,2,3\}, \{1,2\}, \{1,3\}, \{1,2,3\}, \{0,1,2\},$$
$$\{0,1,3\}, \{0,1,2,3\}\}$$
$$\mathcal{P}^*(\{\{0,1\}\}) = \{\{0\}, \{1\}, \{0,1\}\}$$
$$\mathcal{P}^*(\{\{2,3\}\}) = \{\{2\}, \{3\}, \{2,3\}\}$$
$$\mathcal{P}^*(\emptyset) = \{\emptyset\}.$$

---

[12]Equivalently, $\mathcal{P}^*(S)$ is the collection of all sets s obtainable by extracting from each $t \in S$ a non-null $W_t \subseteq t$, and then forming $s = \bigcup_{t \in S} W_t$.

It is easy to check that

$$\Sigma = \Big\{ \, \mathcal{P}^*(\{\{0,1\}, \{2,3\}\}), \ \mathcal{P}^*(\{\{0,1\}\}), \ \mathcal{P}^*(\{\{2,3\}\}), \ \mathcal{P}^*(\emptyset) \, \Big\}$$

is indeed a partition of the powerset of the domain of S, namely of the set $\mathcal{P}(\{0,1,2,3\})$. As the set $\{0,1,2,3\}$ is transitive (in fact, it is the ordinal 4), its powerset is transitive as well. Thus, in particular, Σ is a transitive partition. ∎

Other useful properties of the intersecting powerset operator $\mathcal{P}^*$ are listed in the following lemma.

Lemma 1.31.   *(a) Let g be a* CONTRACTIVE MAP *on X, namely, a map $g\colon X \to \mathcal{P}(\bigcup X)$ such that $g(x) \subseteq x$, for every $x \in X$. Then $\mathcal{P}^*(g[X]) \subseteq \mathcal{P}^*(X)$.*

(b) For every partition Σ and for all $\Gamma_0, \Gamma_1 \subseteq \Sigma$,

$$\mathcal{P}^*(\Gamma_0 \cup \Gamma_1) = \big\{ X \cup Y \mid X \in \mathcal{P}^*(\Gamma_0) \ \wedge \ Y \in \mathcal{P}^*(\Gamma_1) \big\};$$

(c) Let Σ be a partition and $X \subseteq \mathcal{P}^*(\Sigma)$. If $\bigcup \Sigma \setminus \bigcup X \neq \emptyset$, then

$$|\mathcal{P}^*(\Sigma) \setminus X| \geqslant 2^{\sup\{|\sigma| \,|\, \sigma \in \Sigma\} - 1}.$$

(d) Let Σ be a finite partition such that $\bigcup \Sigma$ is infinite and let $u \subseteq \bigcup \Sigma$ be any finite set. Then the set $\{v \in \mathcal{P}^*(\Sigma) : u \subseteq v\}$ is infinite.

Proof. Concerning (a), let g be a contractive map on X, and let $t \in \mathcal{P}^*(g[X])$. Then $t \subseteq \bigcup g[X] \subseteq \bigcup X$. In addition, for every $x \in X$, we have $t \cap g(x) \neq \emptyset$, so that $t \cap x \neq \emptyset$ holds, proving that $t \in \mathcal{P}^*(X)$. Hence $\mathcal{P}^*(g[X]) \subseteq \mathcal{P}^*(X)$.

Concerning (b), let Σ be a partition, $\Gamma_0, \Gamma_1 \subseteq \Sigma$, and $Z \in \mathcal{P}^*(\Gamma_0 \cup \Gamma_1)$. By putting $X := Z \cap \bigcup \Gamma_0$ and $Y := Z \cap \bigcup \Gamma_1$, we plainly have $X \in \mathcal{P}^*(\Gamma_0)$, $Y \in \mathcal{P}^*(\Gamma_1)$, and $Z = X \cup Y$, proving $\mathcal{P}^*(\Gamma_0 \cup \Gamma_1) \subseteq \big\{ X \cup Y \mid X \in \mathcal{P}^*(\Gamma_0) \ \wedge \ Y \in \mathcal{P}^*(\Gamma_1) \big\}$. The converse inclusion is immediate.

To prove (c), let $x_0 \in \bigcup \Sigma \setminus \bigcup X$ and let $\sigma \in \Sigma$. If $x_0 \notin \sigma$, then
$$\big\{ \{x_0\} \cup Y \cup Z \mid Y \in \mathcal{P}^*(\{\sigma\}) \ \wedge \ Z \in \mathcal{P}^*(\Sigma \setminus \{\sigma\}) \big\} \subseteq \mathcal{P}^*(\Sigma) \setminus X,$$
otherwise
$$\big\{ \{x_0\} \cup Y \cup Z \mid Y \in \mathcal{P}(\sigma \setminus \{x_0\}) \ \wedge \ Z \in \mathcal{P}^*(\Sigma \setminus \{\sigma\}) \big\} \subseteq \mathcal{P}^*(\Sigma) \setminus X.$$
In the former case we have $|\mathcal{P}^*(\Sigma) \setminus X| \geqslant 2^{|\sigma|} - 1$, while in the latter we have $|\mathcal{P}^*(\Sigma) \setminus X| \geqslant 2^{|\sigma|-1}$. Since $\sigma \neq \emptyset$, and hence $2^{|\sigma|} - 1 \geqslant 2^{|\sigma|-1}$, in either case we get $|\mathcal{P}^*(\Sigma) \setminus X| \geqslant 2^{|\sigma|-1}$. Since this holds for all $\sigma \in \Sigma$, we conclude that $|\mathcal{P}^*(\Sigma) \setminus X| \geqslant 2^{(\sup\{|\sigma| \,|\, \sigma \in \Sigma\} - 1)}$. [13]

Finally, a verification of (d) runs as follows. Since Σ is finite and $\bigcup \Sigma$ is infinite, the partition Σ must contain at least an infinite block $\overline{\sigma}$. Then, it is an easy matter to check that
$$\big\{ (\textstyle\bigcup \Sigma \setminus \overline{\sigma}) \cup u \cup s \mid s \in \mathcal{P}^*(\overline{\sigma} \setminus u) \big\}$$
is an infinite subset of $\{v \in \mathcal{P}^*(\Sigma) : u \subseteq v\}$, as $\overline{\sigma} \setminus u$ is infinite. □

---

[13] For a cardinal κ, by $(\kappa - 1)$ we intend the cardinal μ such that $\mu + 1 = \kappa$. Notice that if κ is infinite, then $\kappa - 1 = \kappa$.

Remark 1.32. By taking a brief detour off our main course, we mention that the topic of contractive maps (cited in Lemma 1.31(a)) has applications in various fields of research, for instance in *Individual Choice Theory*. Recall that, given a set X of alternatives, a *choice* on X is a contractive map $c \colon \Omega \to \mathrm{pow}(X) \setminus \{\emptyset\}$, where the choice domain $\Omega \subseteq \mathrm{pow}(X) \setminus \{\emptyset\}$ is usually assumed to satisfy suitable properties of closure. Elements of Ω are called *menus*, and elements of a menu are called *items*. Semantically, a choice c associates to each menu $A \in \Omega$ a nonempty subset $c(A) \subseteq A$, which comprises all items of A that are deemed selectable by an economic agent. According to the *Theory of Revealed Preferences*, pioneered by the economist Paul Samuelson (see [Sam38]), a choice is *rationalizable* if it can be retrieved from a binary relation on X by taking all maximal elements of each menu. (Recall that, given a binary relation $\precsim$ on X and a nonempty set $A \subseteq X$, the set of $\precsim$-*maximal* elements of A is given by $\{z \in A : (\forall a \in A)\, (z \precsim a \;\Rightarrow\; a \precsim z)\}$.) In symbols, $c \colon \Omega \to \mathrm{pow}(X) \setminus \{\emptyset\}$ is rationalizable if there exists a binary relation $\precsim$ on X such that the equality $c(A) = \max(A, \precsim)$ holds for any menu $A \in \Omega$. It is well known that rationalizable choices can be characterized by the satisfaction of suitable *axioms of choice consistency*, which are second-order logic formulae codifying rules of coherent selection within menus. For a very recent analysis of rationalizable choices—also in connection with the transitive structure (in the sense of [GW14]) of the rationalizing preference relation—the reader may consult [CGGW16] and references therein. We mention also some very recent satisfiability results for BST-formulae extended with the singleton operator and a choice function symbol subject to various combinations of axioms of choice consistency (see [CGW17]). In Section 2.3, the fragment BST (for Boolean Set Theory) will be examined in depth as a first case study on decidability in set theory. ∎

EXERCISES

Exercise 1.1. *Verify Remark 1.1.*

Exercise 1.2. *Prove Lemma 1.2.*

Exercise 1.3. *Prove that a set is transitive according to Definition 1.5 if and only if it contains as members all the elements which participate in its construction.*

Exercise 1.4. *Prove that each layer $\boldsymbol{V}_\mu$ of the von Neumann standard cumulative hierarchy is transitive, for every ordinal μ.*

Exercise 1.5. *Prove that $\boldsymbol{V}_\omega$ is the collection of all finite sets, whose elements, elements of elements, etc., all are finite.*

Exercise 1.6. *Prove Lemma 1.15.*

Exercise 1.7. *Prove that the transitive closure operator satisfies the following transfinite recursion:*

$$\mathsf{TrCl}(S) = S \cup \bigcup_{s \in S} \mathsf{TrCl}(s)\,.$$

Exercise 1.8. *Prove Lemma 1.17.*

Exercise 1.9. *Given a set S, let $S_0 := S$ and, recursively, $S_{i+1} := \bigcup S_i$, for every natural number $i \in \omega$. Prove that $\mathsf{TrCl}(S) = \bigcup\{S_i \mid i \in \omega\}$.*

Exercise 1.10. *Let Σ be a non-transitive partition. Prove that $\Sigma \cup \{\mathsf{TrCl}(\bigcup\Sigma) \setminus \bigcup\Sigma\}$ is the minimal transitive partition which extends Σ.*

Exercise 1.11. *Prove that the $\mathcal{P}$-completion of a given partition Σ is a $\mathcal{P}$-partition.*

Exercise 1.12. *Prove that if Σ is a $\mathcal{P}$-partition, then there is exactly one block $\overline{\sigma} \in \Sigma$ such that*

$$\bigcup(\Sigma \setminus \{\overline{\sigma}\}) \subseteq \bigcup\bigcup\Sigma \subseteq \mathcal{P}(\bigcup\bigcup\Sigma) = \bigcup\Sigma.$$

Exercise 1.13. *Let Σ^* and σ^* be respectively the $\mathcal{P}$-completion and the external block of a given partition Σ (relative to the $\mathcal{P}$-completion Σ^*). Prove that*

(a) Σ^* is a $\mathcal{P}$-partition;

(b) $\mathcal{P}\big(\mathsf{TrCl}(\bigcup(\Sigma^* \setminus \{\sigma^*\}))\big) = \bigcup\Sigma^*$.

Exercise 1.14. *Complete the proof of Lemma 1.23.*

Exercise 1.15. *Complete the proof of Lemma 1.24.*

Exercise 1.16. *Let $\sim_{\mathcal{F}}$ be the relation over $\bigcup\mathcal{F}$ in Definition 1.25, where $\mathcal{F}$ is a finite collection of sets. Prove that $\sim_{\mathcal{F}}$ is an equivalence relation and that*

$$\left\{ \bigcap\mathcal{E} \setminus \bigcup(\mathcal{F} \setminus \mathcal{E}) \,\middle|\, \emptyset \neq \mathcal{E} \subseteq \mathcal{F} \right\} \setminus \{\emptyset\}$$

is the partition on $\bigcup\mathcal{F}$ induced by it.

Exercise 1.17. *Prove that we have*

$$\mathcal{P}\left(\bigcup S\right) = \bigcup\{\mathcal{P}^*(T) \mid T \subseteq S\},$$

for every family of sets S.

Exercise 1.18. *Let Σ be a $\mathcal{P}$-partition with special block σ^*. Prove that, for every $X \subseteq \Sigma$,*

$$\mathcal{P}^*(X) \subseteq \bigcup\Sigma \quad \Longleftrightarrow \quad \sigma^* \notin X.$$

Chapter 2

The Decision Problem in Set Theory

In connection with the decision problem in set theory, we are interested in specific collections of quantifier-free formulae,[1] referred to as FRAGMENTS or THEORIES throughout the book.[2] These involve various combinations of set-theoretic operators and predicates. In particular, the elementary Boolean set-theoretic operators of binary union $\cup$, binary intersection $\cap$, and set difference $\setminus$, as well as the predicate symbols of equality $=$, set inclusion $\subseteq$, and membership $\in$, will almost always be present. These constructs constitute the *common core*—named MLS—of our theories.[3]

We start the chapter with the presentation of the syntax and semantics of a comprehensive fragment $\mathcal{S}$ of set theory, which encompasses all the theories mentioned in the book. The language of $\mathcal{S}$ is quantifier-free and includes, in addition to the set constructs of MLS, the finiteness predicate symbol $Finite(\cdot)$ and the operators of powerset $\mathcal{P}$, union set $\bigcup$, Cartesian product $\times$, finite enumeration $\{\cdot, \ldots, \cdot\}$, etc. Along with the definition of the satisfiability problem for $\mathcal{S}$ and its variants, we then introduce the novel notion of *satisfiability by partition*. Although equivalent to classical satisfiability by assignments, satisfiability by partition is particularly useful in relation to the decidability techniques that will be developed in the rest of the book. In fact, partitions can be related among them by specific relationships (such as *simulation*, *imitation*, and their variants), enjoying properties of the following form, relative to given fragments $\mathfrak{F}$ of $\mathcal{S}$:

> If a partition $\widehat{\Sigma}$ is related to another partition Σ, then all $\mathfrak{F}$-formulae satisfied by Σ are also satisfied by $\widehat{\Sigma}$.

These relationships are very helpful for proving that certain fragments of $\mathcal{S}$ enjoy a *small model property* (or a *small witness-model property*, in the case of fragments able to force infinite sets), thereby proving their decidability.

To prepare for the much more challenging decision problems that will be examined in the second part of the book, we also provide a detailed presentation of a simple case study concerning the satisfiability problem for the theory BST, for Boolean Set Theory—BST is obtained by dropping from MLS the membership predicate symbol. This constitutes a preliminary example of the *satisfiability-by-partition* approach.

---

[1] For the quantified case, the interested reader may refer to [CFO89, SCO11].

[2] The decision problem for fragments of set theory is the subject matter of the research field of *Computable Set Theory* (cf. [CFO89, COP01, SCO11]).

[3] MLS stands for MULTI-LEVEL SYLLOGISTIC (cf. [FOS80]).

© Springer International Publishing AG, part of Springer Nature 2018
D. Cantone and P. Ursino, *An Introduction to the Technique of Formative Processes in Set Theory*, https://doi.org/10.1007/978-3-319-74778-1_2

We shall also discuss some expressiveness issues, by proving that certain operators and relators, which are not present as primitive symbols in some specific fragments of $\mathcal{S}$, are expressible in such theories anyway. These facts yield two negative model-theoretic results related to decidability and *rank dichotomicity* (the latter is a novel notion strictly related to the small model property).

2.1   The Theory $\mathcal{S}$

2.1.1   Syntax and Semantics of $\mathcal{S}$

We present the syntax and semantics of a comprehensive fragment $\mathcal{S}$ of set theory, and define the decision problem for it (and its subtheories).

The symbols of the language of $\mathcal{S}$ are:

- infinitely many set variables $x, y, z, \ldots$;

- the constant symbol $\varnothing$;

- the set operators $\cdot \cup \cdot, \cdot \cap \cdot, \cdot \setminus \cdot, \cdot \times \cdot, \cdot \otimes \cdot, \{\cdot, \ldots, \cdot\}, \mathcal{P}(\cdot), \bigcup(\cdot), \uplus(\cdot), \bigcap(\cdot);$[4]

- the set predicates $\cdot \subseteq \cdot, \cdot = \cdot, \cdot \in \cdot, Finite(\cdot)$.

The collection of $\mathcal{S}$-TERMS is the smallest set of expressions such that:

- all variables and the constant $\varnothing$ are $\mathcal{S}$-terms;

- if s and t are $\mathcal{S}$-terms, so are $s \cup t$, $s \cap t$, $s \setminus t$, $s \times t$, $s \otimes t$, $\mathcal{P}(s)$, $\bigcup s$, $\uplus s$, and $\bigcap s$;

- if $s_1, \ldots, s_n$ are $\mathcal{S}$-terms, so is $\{s_1, \ldots, s_n\}$.

$\mathcal{S}$-ATOMS have the form

$$s \subseteq t, \qquad s = t, \qquad s \in t, \qquad Finite(s),$$

where s, t are $\mathcal{S}$-terms.

$\mathcal{S}$-FORMULAE are propositional combinations of $\mathcal{S}$-atoms, by means of the usual logical connectives $\wedge$ (conjunction), $\vee$ (disjunction), $\neg$ (negation), $\rightarrow$ (implication), $\leftrightarrow$ (bi-implication), etc. $\mathcal{S}$-LITERALS are $\mathcal{S}$-atoms and their negations.

An $\mathcal{S}$-EXPRESSION is either an $\mathcal{S}$-term or an $\mathcal{S}$-formula.

For an $\mathcal{S}$-formula or an $\mathcal{S}$-term Φ, we denote by $\mathrm{Vars}(\Phi)$ the collection of the set variables occurring in Φ. Thus, for instance,

$$\mathrm{Vars}\big(\{x \cup \varnothing, \mathcal{P}\big(y \cap (z \setminus \{x\})\big)\} \setminus \bigcup(x \cap y)\big) = \{x, y, z\}.$$

To define the size of $\mathcal{S}$-expressions, it is useful to introduce the notion of *syntax tree*.

Definition 2.1 ($\mathcal{S}$-syntax tree). *The* SYNTAX TREE *of an $\mathcal{S}$-expression (or $\mathcal{S}$-syntax tree) is a labeled ordered tree. $\mathcal{S}$-syntax trees can be recursively defined by the following rules (for simplicity, our definition will cover only the case of $\mathcal{S}$-terms; but it is an easy matter to generalize it also to $\mathcal{S}$-formulae—the reader is asked to do so in Exercise 2.1):*

---

[4]We recall that by $\otimes$ and $\uplus$ we denote, respectively, the operators *unordered Cartesian product* and *disjoint union set*.

(i) the syntax tree of the constant $\varnothing$, or of a set variable x, is the one-node tree whose root is labeled by $\varnothing$, or by x, respectively;

(ii) the syntax tree of an $\mathcal{S}$-term of the form $\mathsf{U}\,s$, where U stands for any unary set operator such as $\mathcal{P}(\cdot)$, $\bigcup(\cdot)$, $\biguplus(\cdot)$, and $\bigcap(\cdot)$, is a tree whose root, labeled by U, has a unique child, pointing to the root of the syntax tree of s;

(iii) the syntax tree of an $\mathcal{S}$-term of the form $s\,\mathsf{B}\,t$, where B stands for any binary set operator such as $\cdot\cup\cdot$, $\cdot\cap\cdot$, $\cdot\setminus\cdot$, $\cdot\times\cdot$, and $\cdot\otimes\cdot$, is a tree whose root, labeled by B, has exactly two children: the left child points to the root of the syntax tree of s, and the right child points to the root of the syntax tree of t;

(iv) the syntax tree of an $\mathcal{S}$-term of the form $\{s_1,\ldots,s_n\}$, with $n \geqslant 1$, is a tree whose root, labeled by $\{\cdot\}$, has exactly n children, where the i-th child points to the root of the syntax tree of s_i, for $i = 1,\ldots,n$. ■

Figure 2.1 shows the syntax tree of the $\mathcal{S}$-term

$$\big\{x \cup \varnothing, \mathcal{P}\big(y \cap (z \setminus \{x\})\big)\big\} \setminus \textstyle\bigcup(x \cap y). \tag{2.1}$$

The SIZE or LENGTH of Φ, written $|\Phi|$, is then defined as the number of nodes in the syntax tree of Φ. For instance, the size of the $\mathcal{S}$-term (2.1) is 16.

$\big\{x \cup \varnothing, \mathcal{P}\big(y \cap (z \setminus \{x\})\big)\big\} \setminus \bigcup(x \cap y)$	$\mathcal{S}$-term
$\neg\big(z \cup x \in \big\{x \cup \varnothing, \mathcal{P}\big(y \cap (z \setminus \{x\})\big)\big\} \to (x \notin \bigcap z \ \vee \ z \in \bigcup x)\big)$	$\mathcal{S}$-formula
$\mathrm{Vars}\Big(\neg\big(z \cup x \in \big\{x \cup \varnothing, \mathcal{P}\big(y \cap (z \setminus \{x\})\big)\big\} \to (x \notin \bigcap z \ \vee \ z \in \bigcup x)\big)\Big) = \{x,y,z\}$	

Table 2.1: A few examples: an $\mathcal{S}$-term, an $\mathcal{S}$-formula, and the map $\mathrm{Vars}(\cdot)$.

The semantics of $\mathcal{S}$ is defined in the most natural way.

Definition 2.2 (Set assignments). *A* SET ASSIGNMENT *M is any map from a collection V of set variables (called the* VARIABLES DOMAIN OF *M) into the universe $\boldsymbol{\mathcal{V}}$ of all sets (in short, $M \in \boldsymbol{\mathcal{V}}^V$ or $M \in \{\text{sets}\}^V$). The* SET DOMAIN OF *M is the set $\bigcup M[V] = \bigcup_{v \in V} Mv$, and the* RANK OF *$M$ is the rank of its set domain, namely,*

$$\mathsf{rk}\,M \ := \ \mathsf{rk}(\textstyle\bigcup M[V]).$$

A set assignment M is FINITE *(resp.,* HEREDITARILY FINITE*), if so is its set domain.* ■

For a set assignment M over a collection V of set variables, when V is finite, $\mathsf{rk}\,M = \max_{v \in V} \mathsf{rk}\,Mv$.

For any set operator, we shall use the same symbol to denote both the operator and its standard interpretation, as in the following definition. This will result in increased readability while not introducing any ambiguity.

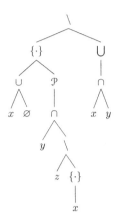

Figure 2.1: Syntax tree of the $\mathcal{S}$-term $\{x \cup \varnothing, \mathcal{P}(y \cap (z \setminus \{x\}))\} \setminus \bigcup(x \cap y)$. Its total number of nodes is 16.

Definition 2.3 (Interpretation of set operators). *Let M be a set assignment and V its variables domain. Also, let $s, t, s_1, \ldots, s_n$ be $\mathcal{S}$-terms whose variables occur in V. We put, recursively,*

$$M\varnothing := \emptyset$$
$$M(s \cup t) := Ms \cup Mt$$
$$M(s \cap t) := Ms \cap Mt$$
$$M(s \setminus t) := Ms \setminus Mt$$
$$M(\mathcal{P}(s)) := \mathcal{P}(Ms) := \{u \mid u \subseteq Ms\}$$
$$M(\textstyle\bigcup s) := \textstyle\bigcup Ms := \{u \mid u \in u' \in Ms, \text{ for some } u'\}$$
$$M(\uplus s) := \uplus Ms := \{u \mid u \in u' \in Ms, \text{ for a unique } u'\}$$
$$M(\textstyle\bigcap s) := \textstyle\bigcap Ms := \{u \mid u \in u', \text{ for every } u' \in Ms\} \quad \text{(provided that } Ms \neq \emptyset)$$
$$M(s \times t) := Ms \times Mt := \{\{\{u\}, \{u, u'\}\} \mid u \in Ms, u' \in Mt\}$$
$$M(s \otimes t) := Ms \otimes Mt := \{\{u, u'\} \mid u \in Ms, u' \in Mt\}. \qquad \blacksquare$$

Thus, $Ms \times Mt$ is the Cartesian product of Ms and Mt, namely, the collection of the ordered pairs *à la* Kuratowski formed with the elements of Ms (first elements) and of Mt (second elements). Further, $Ms \otimes Mt$ is the unordered Cartesian product of Ms and Mt, namely, the collection of the (unordered) pairs formed with the elements of Ms and of Mt.

Definition 2.4 (Interpretation of set relators). *For all $\mathcal{S}$-terms s, t such that $\mathrm{Vars}(s)$,*

$\mathrm{Vars}(t) \subseteq V$, *we put:*

$$(s \in t)^M := \begin{cases} \mathbf{t} & \textit{if } Ms \in Mt \\ \mathbf{f} & \textit{otherwise} \end{cases} \qquad (s = t)^M := \begin{cases} \mathbf{t} & \textit{if } Ms = Mt \\ \mathbf{f} & \textit{otherwise} \end{cases}$$

$$(s \subseteq t)^M := \begin{cases} \mathbf{t} & \textit{if } Ms \subseteq Mt \\ \mathbf{f} & \textit{otherwise} \end{cases} \qquad (Finite(s))^M := \begin{cases} \mathbf{t} & \textit{if } Ms \textit{ is finite} \\ \mathbf{f} & \textit{otherwise,} \end{cases}$$

where $\mathbf{t}$ *and* $\mathbf{f}$ *stand the truth-values* true *and* false, *respectively.* ∎

Propositional connectives are interpreted in the standard way.

Definition 2.5 (Interpretation of propositional connectives). *For all* $\mathcal{S}$-*formulae* Φ, Ψ *such that* $\mathrm{Vars}(\Phi), \mathrm{Vars}(\Psi) \subseteq V$, *we put recursively*

$$(\Phi \wedge \Psi)^M := \Phi^M \wedge \Psi^M, \qquad (\Phi \vee \Psi)^M := \Phi^M \vee \Psi^M,$$
$$(\Phi \rightarrow \Psi)^M := \Phi^M \rightarrow \Psi^M, \qquad (\neg \Phi)^M := \neg(\Phi^M), \qquad \textit{etc.} \quad ∎$$

Next we introduce some common terminology related to the satisfiability relation.

Definition 2.6. *A set assignment* M *is said to* SATISFY *an* $\mathcal{S}$-*formula* Φ *if* $\Phi^M = \mathbf{t}$ *holds, in which case we also write* $M \models \Phi$, *and say that* M *is a* MODEL *for* Φ. *If* Φ *has a model, we say that* Φ *is* SATISFIABLE; *otherwise, we say that* Φ *is* UNSATISFIABLE. *In particular, if* Φ *has a finite model (resp., a hereditarily finite model), we say that it is* FINITELY SATISFIABLE *(resp.,* HEREDITARILY FINITELY SATISFIABLE*). Furthermore, if* Φ *has a model* M *such that* $Mx \neq My$ *for all distinct variables* $x, y \in \mathrm{Vars}(\Phi)$, *we say that it is* INJECTIVELY SATISFIABLE. *If* $M' \models \Phi$ *for every set assignment* M' *defined over* $\mathrm{Vars}(\Phi)$, *then* Φ *is said to be* TRUE, *and we write* $\models \Phi$. *Finally, two* $\mathcal{S}$-*formulae* Φ *and* Ψ *are said to be* EQUISATISFIABLE *if* Φ *is satisfiable if and only if so is* Ψ. ∎

Example 2.7. Let $\Phi \overset{\mathrm{Def}}{:=} x \in y \wedge y \in z \wedge z \setminus x \neq \varnothing$. The assignment M defined over $\mathrm{Vars}(\Phi)$ by

$$Mx := \varnothing, \qquad My := \{\varnothing\}, \qquad Mz := \{\{\varnothing\}\}$$

clearly satisfies Φ. ∎

Example 2.8. The formula

$$x \in y \wedge y \in z \wedge z \setminus x = \varnothing \tag{2.2}$$

is unsatisfiable. Indeed, since $z \setminus x = \varnothing$ is equivalent to $z \subseteq x$, then (2.2) yields

$$x \in y \wedge y \in x,$$

which contradicts the well-foundedness of $\in$. Therefore the formula

$$(x \in y \wedge y \in z) \rightarrow z \setminus x \neq \varnothing$$

is true (regardless of the assignment). Instead, the formula

$$(x \in y \land y \in z) \to z \setminus x = \varnothing$$

is not true, since the formula

$$x \in y \land y \in z \land z \setminus x \neq \varnothing$$

is satisfiable (see Example 2.7). ∎

Example 2.9. The formula

$$\Phi_1 \overset{\text{Def}}{:=} x = \bigcup x \land x \neq \varnothing$$

is satisfied by the infinite model M over $\{x\}$ such that

$$Mx := \{\emptyset^n \mid n \in \mathbb{N}\},$$

where we are using the nesting notation $\emptyset^n$ introduced in Example 1.12.

In addition, Φ admits only infinite models (therefore it is not finitely satisfiable). Indeed, if $M \models \Phi$, then $\emptyset \neq Mx = \bigcup Mx$. If the model M were to be finite, then the set Mx would contain a member s of maximal rank. It would then follow that $s \in \bigcup Mx$, hence $s \in t$ for some $t \in Mx$. Thus, we can conclude that $\text{rk } s < \text{rk } t$, contradicting the maximality of the rank of s in Mx.

Likewise, the formulae

$$\Phi_2 \overset{\text{Def}}{:=} x \times x \subseteq x \land x \neq \varnothing$$

$$\Phi_3 \overset{\text{Def}}{:=} x \otimes x \subseteq x \land x \neq \varnothing$$

admit only infinite models (the reader is asked to prove it in Exercise 2.3). ∎

2.1.2   The Decision Problem for Subtheories of $\mathcal{S}$

We define the satisfiability problem and its related terminology.

Definition 2.10. *Let $\mathfrak{F}$ be any subtheory of $\mathcal{S}$, so that the formulae of $\mathfrak{F}$ form a subset of the collection of all the formulae of $\mathcal{S}$. The* DECISION PROBLEM *(or* SATISFIABILITY PROBLEM, *or* SATISFACTION PROBLEM*) for $\mathfrak{F}$ is the problem of establishing algorithmically whether any given $\mathfrak{F}$-formula is satisfiable. If the decision problem for $\mathfrak{F}$ is solvable, then $\mathfrak{F}$ is said to be* DECIDABLE. *A* DECISION PROCEDURE *(or* SATISFIABILITY TEST*) for $\mathfrak{F}$ is any algorithm that solves the decision problem for $\mathfrak{F}$. The* FINITE SATISFIABILITY PROBLEM *for $\mathfrak{F}$ is the problem of establishing algorithmically whether any given $\mathfrak{F}$-formula is finitely satisfiable. The* INJECTIVE SATISFIABILITY PROBLEM *for $\mathfrak{F}$ is the problem of establishing algorithmically whether any given $\mathfrak{F}$-formula is injectively satisfiable.* ∎

By making use of disjunctive normal form, the satisfiability problem for $\mathcal{S}$ can be readily reduced to the same problem for conjunctions of $\mathcal{S}$-literals. In addition, by suitably

---

(s1) a literal of type $x = y$ can be replaced by the equivalent literal $x = y \cup y$,

(s2) a literal of type $x \nsubseteq y$ is equisatisfiable with $z' = x \setminus y \wedge z' \neq \varnothing$,

(s3) the constant $\varnothing$ can be eliminated by replacing it with a new variable $y_\varnothing$ and adding the conjunct $y_\varnothing = y_\varnothing \setminus y_\varnothing$,

(s4) a literal of type $x = y \cap z$ is equisatisfiable with $y' = y \setminus z \wedge x = y \setminus y'$,

(s5) a literal of type $x \subseteq y$ is equisatisfiable with $y = x \cup y$,

(s6) a literal of type $x \neq y$ is equisatisfiable with $x \in z' \wedge y \notin z'$,[5]

(s7) a literal of type $x \notin y$ is equisatisfiable with $x \in z' \wedge z' = z' \setminus y$.

---

y' and z' stand for fresh set variables.

Table 2.2: Simplification rules for literals in (2.3).

introducing fresh set variables to name subterms of the forms

$$t_1 \cup t_2, \quad t_1 \cap t_2, \quad t_1 \setminus t_2, \quad \mathcal{P}(t), \quad \bigcup t, \quad \bigcap t, \quad t_1 \times t_2, \quad t_1 \otimes t_2,$$

where t_1, t_2, t are S-terms, the satisfiability problem for S can further be reduced to the satisfiability problem for conjunctions of S-literals of the following types:

$$
\begin{array}{llll}
x = y \cup z, & x = y \cap z, & x = y \setminus z, & x = \{y_1, \dots, y_H\}, \\
x = \mathcal{P}(y), & x = \bigcup y, & x = \biguplus y, & x = \bigcap y, \\
x = y, & x \neq y, & x = y \times z, & x = y \otimes z, \\
x \in y, & x \notin y, & x \subseteq y, & x \nsubseteq y,
\end{array}
\tag{2.3}
$$

where $x, y, z, y_1, \dots, y_H$ stand for set variables or for the constant $\varnothing$.

Finally, by applying the simplification rules (s1)–(s7) in Table 2.2, some of the literals in (2.3) can be expressed in terms of the remaining ones. As a consequence, the satisfiability problem for S can be reduced to the satisfiability problem for conjunctions of S-atoms of the following types:

$$
\begin{array}{llllll}
x = y \cup z, & x = y \setminus z, & x \in y, & x = \{y_1, \dots, y_H\}, & x = y \otimes z, \\
x = y \times z, & x = \bigcup y, & x = \biguplus y, & x = \bigcap y, & x = \mathcal{P}(y),
\end{array}
\tag{2.4}
$$

where $x, y, y_1, \dots, y_H$ stand for variables. These are called NORMALIZED CONJUNCTIONS of S. Plainly, working with normalized conjunctions simplifies the completeness and correctness proofs of decision procedures.

Although the reduction technique of normalized conjunctions has been illustrated in the case of the whole theory S, it can be easily adapted to subtheories of S as well.

Remark 2.11. The additional two simplification rules (s8) and (s9) in Table 2.3 would theoretically allow us to further simplify (2.4) by getting rid of literals of type $x \in y$ and

---

(s8) a literal of type $x \in y$ is equisatisfiable with $z' = \{x\} \wedge y = y \cup z'$,

(s9) a literal of type $x = \{y_1, \ldots, y_H\}$ is equisatisfiable with the conjunction

$$x_1' = \{y_1\} \wedge \ldots \wedge x_H' = \{y_H\} \wedge x = x_1' \cup \cdots \cup x_H' \,.$$

$x_1', \ldots, x_H', z'$ stand for fresh set variables.

Table 2.3: Additional simplification rules for literals of type $x \in y$ and $x = \{y_1, \ldots, y_H\}$.

by replacing the literals of the form $x = \{y_1, \ldots, y_H\}$ with simpler singleton literals of the form $x = \{y\}$. However, we prefer to conform to current literature, thus avoid making any further simplification.

Notice that, for $H > 2$, the literal $x = x_1' \cup \cdots \cup x_H'$ in rule (s9) can be expressed by the equisatisfiable conjunction

$$x = x_1' \cup z_1 \wedge z_1 = x_2' \cup z_2 \wedge \ldots \wedge z_{H-3} = x_{H-2}' \cup z_{H-2} \wedge z_{H-2} = x_{H-1}' \cup x_H' \,,$$

where the z_i's are intended to be fresh set variables. ∎

We observe that a satisfiability test for a subtheory $\mathfrak{F}$ of $\mathcal{S}$ can also be used to decide whether any given formula Φ in $\mathfrak{F}$ is TRUE. In fact, a formula Φ is true if and only if its negation $\neg\Phi$ is unsatisfiable.

Decidable fragments of set theory

Over the years, several subtheories of $\mathcal{S}$ have been shown to have a decidable satisfiability problem. We mention, in particular, the theory Multi-Level Syllogistic (MLS), which is the common kernel of most decidable fragments of set theory investigated in the field of computable set theory. Specifically, MLS is the propositional combination of atomic formulae of the following three types[6]

$$x = y \cup z, \qquad x = y \setminus z, \qquad x \in y. \qquad (2.5)$$

Below we list some of the subtheories of $\mathcal{S}$ whose decision problem has been solved (after each acronym, we indicate the operators and relators admitted in the theory, and some references to the literature):

---

[6]As remarked above, intersection and set inclusion are easily expressible by means of literals of type (2.5).

MLS:	$\cup, \cap, \setminus, \subseteq, =, \in$	(cf. [FOS80])
MLSS:	$\cup, \cap, \setminus, \subseteq, =, \in, \{\cdot\}$	(cf. [FOS80])
MLSU:	$\cup, \cap, \setminus, \subseteq, =, \in, \bigcup(\cdot)$	(cf. [CFS87])
MLSSI:	$\cup, \cap, \setminus, \subseteq, =, \in, \{\cdot\}, \bigcap(\cdot)$	(cf. [CC91])[7]
MLSP:	$\cup, \cap, \setminus, \subseteq, =, \in, \mathcal{P}(\cdot)$	(cf. [CFS85])
MLSSP:	$\cup, \cap, \setminus, \subseteq, =, \in, \{\cdot\}, \mathcal{P}(\cdot)$	(cf. [Can91] and [COU02])
MLSSPF:	$\cup, \cap, \setminus, \subseteq, =, \in, \{\cdot\}, \mathcal{P}(\cdot), \mathit{Finite}(\cdot)$	(cf. [CU14])
$\cdots$	$\cdots$	$\cdots$

The interested reader can find an extensive treatment of these results in [CFO89] and [COP01]. In this book we present in some detail the proofs of the decidability for MLSSP and MLSSPF (in Chapters 4 and 5, respectively).

Since the theories MLSP, MLSSP, and MLSSPF are the main subjects of this book, for convenience we describe explicitly their normalized conjunctive forms.

Arguing as in Section 2.1.2, it is easy to show that the satisfiability problem for MLSP can be reduced to the satisfiability problem for conjunctions of literals of the following types:

$$ x = y \cup z, \qquad x = y \setminus z, \qquad x \in y, \qquad x \notin y, \quad x = \mathcal{P}(y), $$

where x, y, z stand for set variables. These will be called NORMALIZED CONJUNCTIONS OF MLSP.

Likewise, the satisfiability problem for MLSSP can be reduced to the satisfiability problem for conjunctions of literals of the following types:

$$ x = y \cup z, \quad x = y \setminus z, \quad x \neq y, \quad x \in y, \quad x \notin y, \quad x = \mathcal{P}(y), \quad x = \{y_1, \ldots, y_H\}, \quad (2.6) $$

where $x, y, z, y_1, \ldots, y_H$ stand for set variables. These will be called NORMALIZED CONJUNCTIONS OF MLSSP. In addition, for a fixed $L \geqslant 0$, a NORMALIZED L-BOUNDED MLSSP-CONJUNCTION is a conjunction of $\mathcal{S}$-literals of the types (2.6) such that, for every conjunct of the form $x = \{y_1, \ldots, y_H\}$, the condition $0 \leqslant H \leqslant L$ must be satisfied.

Finally, the NORMALIZED CONJUNCTIONS OF MLSSPF are conjunctions of literals of the following types:

$$ \begin{aligned} &x = y \cup z, & &x = y \setminus z, & &x \in y, & &x \notin y, \\ &x = \mathcal{P}(y), & &x = \{y_1, \ldots, y_H\}, & &x = \mathit{Finite}(y), & & \end{aligned} \qquad (2.7) $$

where, again, $x, y, z, y_1, \ldots, y_H$ stand for set variables.

Other theories of interest to the developments that will take place in Sections 2.4 and 2.5 are:

- MLSC, the extension of MLS with literals of the type $x = y \times z$,

---

[7]In fact, [CC91] solves the decision problem for the extension of MLSSI with strict and non-strict rank comparison literals, namely, with literals of the form $x < y$ and $x \leqslant y$, whose semantics is

$$ (s < t)^M := \begin{cases} \mathbf{t} & \text{if rk } Ms < \text{rk } Mt \\ \mathbf{f} & \text{otherwise} \end{cases}, \qquad (s \leqslant t)^M := \begin{cases} \mathbf{t} & \text{if rk } Ms \leqslant \text{rk } Mt \\ \mathbf{f} & \text{otherwise} \end{cases}. $$

- MLSuC, the extension of MLS with literals of the type $x = y \otimes z$, and

- MLSuC$^+$, the extension of MLS with literals of the types $x = y \otimes z$ and $x = \uplus y$.[8]

Remark 2.12. The decision problem for the theories MLSC and MLSuC is still open, whereas, as will be shown in Section 2.5.1, the satisfiability problem for MLSuC$^+$ is undecidable. We expect that a variation of the technique of formative processes, which will be reviewed in Chapter 3, will allow us to settle the decision problem for MLSC and MLSuC. ∎

In several cases, decidability has been shown by proving a *small model property*. This is the case, for instance, for MLSP and MLSSP.

Small model property for subtheories of $\mathcal{S}$

Definition 2.13 (Small model property). *A subtheory $\mathfrak{F}$ of $\mathcal{S}$ enjoys the* SMALL MODEL PROPERTY *if there exists a computable function[9] $r_{\mathfrak{F}} \colon \mathbb{N} \to \mathbb{N}$ such that any satisfiable formula Φ of $\mathfrak{F}$ is satisfied by some (finite) model whose rank is bounded by $r_{\mathfrak{F}}(|\Phi|)$ (we recall that $|\Phi|$ is size of Φ). We shall refer to the function $r_{\mathfrak{F}}$ as the* RANK-BOUND FUNCTION *for $\mathfrak{F}$.* ∎

Observe that, if a subtheory $\mathfrak{F}$ of $\mathcal{S}$ enjoys the small model property, then it is decidable. Indeed, a satisfiability test for $\mathfrak{F}$ could run as follows. Let $r_{\mathfrak{F}}$ be a rank-bound function for $\mathfrak{F}$, and Φ any normalized conjunction of $\mathfrak{F}$. The set assignments over Vars(Φ) whose rank is bounded by $r_{\mathfrak{F}}(|\Phi|)$ are finitely many, and can be effectively generated. In addition, for each of them it can be effectively tested whether it satisfies all the conjuncts of Φ. If no set assignment of bounded rank that satisfies Φ is found, then, by the rank-boundedness property, Φ must be unsatisfiable; otherwise, Φ is satisfiable.

Further, we observe that, for a subtheory $\mathfrak{F}$ enjoying the small model property, the satisfiability problem and the hereditarily finite satisfiability problem are plainly equivalent.

As already remarked, the decidability of MLSP and MLSSP has been obtained by showing that they enjoy the small model property, whereas the decision problem for MLSSPF has been reduced to that of MLSSP via a small witness-model property.

Venn partitions of set assignments

The definition of Venn partition (and its variants) for a collection of sets outlined in Chapter 1 can be readily adapted to set assignments.

Definition 2.14. *Given a set assignment M over a finite collection V of variables, the* VENN PARTITION Σ^{V}_M INDUCED BY M *is the Venn partition $\Sigma^{\mathsf{V}}_{\mathcal{F}_M}$ of the collection of sets $\mathcal{F}_M := \{Mv \mid v \in V\}$.* ∎

---

[8]We recall again that $\otimes$ and $\uplus$ are the unordered Cartesian product and the disjoint unary operators, respectively.

[9]Roughly speaking, a (possibly partial) function $c \colon \omega \nrightarrow \omega$ is *computable* when there is an algorithm that, on any input $n \in \mathsf{dom}(c)$, calculates the value $c(n)$ in a finite number of steps; otherwise, it performs an endless computation; see [Rog67].

Thus,

$$\Sigma_M^{\mathsf{V}} := \Big\{ \bigcap_{u \in U} Mu \setminus \bigcup_{v \in V \setminus U} Mv \ \Big| \ \emptyset \neq U \subseteq V \Big\} \setminus \{\emptyset\}$$

is the (proper) Venn partition induced by M.

Definition 2.15. *The* TRANSITIVE VENN PARTITION Σ_M^{T} INDUCED BY M *is the transitive Venn partition* $\Sigma_{\mathcal{F}_M}^{\mathsf{T}}$ *of the collection of sets* $\mathcal{F}_M$. *Hence,*

$$\Sigma_M^{\mathsf{T}} := \begin{cases} \Sigma_M^{\mathsf{V}} & \text{if } \bigcup \Sigma_M^{\mathsf{V}} \text{ is transitive} \\ \Sigma_M^{\mathsf{V}} \cup \{ \mathsf{TrCl}(\bigcup \mathcal{F}_M) \setminus \bigcup \mathcal{F}_M \} & \text{otherwise.} \end{cases}$$ ∎

Definition 2.16. *A* VENN $\mathcal{P}$-PARTITION $\Sigma_M^{\mathcal{P}}$ INDUCED BY M *is any Venn $\mathcal{P}$-partition* $\Sigma_{\mathcal{F}_M}^{\mathcal{P}}$ *of the collection of sets* $\mathcal{F}_M$. ∎

2.1.3 Satisfiability by Partitions

Let V be a finite collection of set variables and Σ a finite partition. Also, let $\mathfrak{I} \colon V \to \mathcal{P}(\Sigma)$, called a PARTITION ASSIGNMENT. The map $\mathfrak{I}$ induces in a natural way a set assignment $M_{\mathfrak{I}}$ over V by putting

$$M_{\mathfrak{I}} v := \bigcup \mathfrak{I}(v), \qquad \text{for } v \in V.$$

Definition 2.17 (Satisfiability by partition). *Let* $\mathfrak{I} \colon V \to \mathcal{P}(\Sigma)$ *be a partition assignment over a finite collection V of set variables, with Σ a finite partition. Also, let Φ be an $\mathcal{S}$-formula such that* $\mathrm{Vars}(\Phi) \subseteq V$. *We say that the partition Σ* SATISFIES Φ VIA THE MAP $\mathfrak{I}$, *and write* $\Sigma / \mathfrak{I} \models \Phi$, *if the set assignment $M_{\mathfrak{I}}$ induced by $\mathfrak{I}$ satisfies Φ. We say that Σ* SATISFIES Φ, *and write* $\Sigma \models \Phi$, *if Σ satisfies Φ via some map* $\mathfrak{I} \colon V \to \mathcal{P}(\Sigma)$.

When Σ is a finite $\mathcal{P}$-partition with special place σ^* we say that Σ STRONGLY SATISFIES Φ, *and write* $\Sigma \models^* \Phi$, *if Σ satisfies Φ via some map* $\mathfrak{I} \colon V \to \mathcal{P}(\Sigma \setminus \{\sigma^*\})$, *namely, such that $\mathfrak{I}(v)$ is a collection of internal blocks, for $v \in V$.*

Thus, if an $\mathcal{S}$-formula Φ is satisfied by some partition, then it is satisfied by some set assignment. It is easy to see that the converse holds as well; the reader is asked to prove this fact in Exercise 2.5.

As a consequence, the notions of satisfiability by set assignments and that of satisfiability by partitions coincide, as formally stated in the following lemma:

Lemma 2.18. *Let Φ be an $\mathcal{S}$-formula. Then $M \models \Phi$, for some set assignment M, if and only if $\Sigma \models \Phi$, for some partition Σ.*

The following lemma shows that certain properties of the set assignment $M_{\mathfrak{I}}$ induced by a given map $\mathfrak{I} \colon V \to \mathcal{P}(\Sigma)$ are reflected on the map $\mathfrak{I}$.

Lemma 2.19. *Let Σ be a partition, V a finite collection of set variables, and $\mathfrak{I} \colon V \to \mathcal{P}(\Sigma)$ a given partition assignment. Then, for $x, y, z \in V$, we have*

(a) $M_{\mathfrak{I}} \models x = y \cup z$ *if and only if* $\mathfrak{I}(x) = \mathfrak{I}(y) \cup \mathfrak{I}(z)$,

(b) $M_{\mathfrak{I}} \models x = y \setminus z$ *if and only if* $\mathfrak{I}(x) = \mathfrak{I}(y) \setminus \mathfrak{I}(z)$, *(c)* $M_{\mathfrak{I}} \models x \neq y$ *if and only if* $\mathfrak{I}(x) \neq \mathfrak{I}(y)$,

where $M_{\mathfrak{I}}$ is the set assignment induced by $\mathfrak{I}$.

Proof. Concerning (a), let us first assume that $M_{\mathfrak{I}} \models x = y \cup z$. Hence, $\bigcup \mathfrak{I}(x) = \bigcup \mathfrak{I}(y) \cup \bigcup \mathfrak{I}(z)$. Let $\sigma \in \Sigma$; then we have:

$$
\begin{aligned}
\sigma \in \mathfrak{I}(x) \quad &\Longleftrightarrow \quad \sigma \subseteq \bigcup \mathfrak{I}(x) && \text{(by Lemma 1.23(b))} \\
&\Longleftrightarrow \quad \sigma \subseteq \bigcup \mathfrak{I}(y) \cup \bigcup \mathfrak{I}(z) \\
&\Longleftrightarrow \quad \sigma \cap \bigcup \mathfrak{I}(y) \neq \emptyset \ \vee \ \sigma \cap \bigcup \mathfrak{I}(z) \neq \emptyset && \text{(by Lemma 1.23(a),(b))} \\
&\Longleftrightarrow \quad \sigma \in \mathfrak{I}(y) \ \vee \ \sigma \in \mathfrak{I}(z) && \text{(by Lemma 1.23(a))} \\
&\Longleftrightarrow \quad \sigma \in \mathfrak{I}(y) \cup \mathfrak{I}(z),
\end{aligned}
$$

so that $\mathfrak{I}(x) = \mathfrak{I}(y) \cup \mathfrak{I}(z)$.

For the converse, if $\mathfrak{I}(x) = \mathfrak{I}(y) \cup \mathfrak{I}(z)$, then we have

$$
M_{\mathfrak{I}} x = \bigcup \mathfrak{I}(x) = \bigcup \big(\mathfrak{I}(y) \cup \mathfrak{I}(z)\big) = \bigcup \mathfrak{I}(y) \cup \bigcup \mathfrak{I}(z) = M_{\mathfrak{I}} y \cup M_{\mathfrak{I}} z,
$$

i.e., $M_{\mathfrak{I}} \models x = y \cup z$, proving (a).

Concerning (b), much as in the previous case it can be shown that if $M_{\mathfrak{I}} \models x = y \setminus z$ then $\mathfrak{I}(x) = \mathfrak{I}(y) \setminus \mathfrak{I}(z)$.

For the converse, let us assume that $\mathfrak{I}(x) = \mathfrak{I}(y) \setminus \mathfrak{I}(z)$ and let $s \in M_{\mathfrak{I}} x$. Then, for some $\sigma \in \Sigma$, we have $s \in \sigma \in \mathfrak{I}(x)$. But then

$$
\begin{aligned}
\sigma \in \mathfrak{I}(x) \quad &\Longrightarrow \quad \sigma \in \mathfrak{I}(y) \setminus \mathfrak{I}(z) \\
&\Longrightarrow \quad \sigma \subseteq \bigcup \mathfrak{I}(y) \ \wedge \ \sigma \cap \bigcup \mathfrak{I}(z) = \emptyset && \text{(by Lemma 1.23(a))} \\
&\Longrightarrow \quad \sigma \subseteq \bigcup \mathfrak{I}(y) \setminus \bigcup \mathfrak{I}(z) \\
&\Longrightarrow \quad \sigma \subseteq M_{\mathfrak{I}} y \setminus M_{\mathfrak{I}} z.
\end{aligned}
$$

Hence, $s \in M_{\mathfrak{I}} y \setminus M_{\mathfrak{I}} z$ and therefore $M_{\mathfrak{I}} x \subseteq M_{\mathfrak{I}} y \setminus M_{\mathfrak{I}} z$. Similarly, one can prove the converse inclusion $M_{\mathfrak{I}} y \setminus M_{\mathfrak{I}} z \subseteq M_{\mathfrak{I}} x$, and therefore $M_{\mathfrak{I}} x = M_{\mathfrak{I}} y \setminus M_{\mathfrak{I}} z$ holds, completing the proof of (b).

Finally, concerning (c), we have

$$
M_{\mathfrak{I}} \models x \neq y \iff M_{\mathfrak{I}} \nvDash x = y \cup y \overset{\text{by (a)}}{\iff} \mathfrak{I}(x) \neq \mathfrak{I}(y) \cup \mathfrak{I}(y) \iff \mathfrak{I}(x) \neq \mathfrak{I}(y). \qquad \square
$$

By a suitable analysis of the partitions satisfying the formulae of a given fragment $\mathfrak{F}$ of $\mathcal{S}$, it is possible to gain insight in the satisfiability problem for $\mathfrak{F}$, which in some cases may lead to a decision procedure. The latter point is illustrated in the following section.

2.2   Relating Partitions

For decidability purposes, partitions can be related to each other via suitably characterized maps that depend on the fragments of interest.

Definition 2.20 (Image-map). *Let Σ and $\widehat{\Sigma}$ be partitions. Given $\beta \colon \Sigma \to \widehat{\Sigma}$, we put*

$$
\overline{\beta}(\Gamma) := \beta[\Gamma]
$$

for $\Gamma \subseteq \Sigma$, and call the map $\overline{\beta}\colon \mathcal{P}(\Sigma) \to \mathcal{P}(\widehat{\Sigma})$ the IMAGE-MAP RELATED TO β. ■

As stated in the following lemma, if β is injective, then so is the image-map $\overline{\beta}$ related to it; in addition, $\overline{\beta}$ distributes over union and set difference.

Lemma 2.21. *Let $\beta\colon \Sigma \rightarrowtail \widehat{\Sigma}$ be an injective map, where Σ and $\widehat{\Sigma}$ are partitions. Then, for all $\Gamma, \Gamma_1, \Gamma_2 \subseteq \Sigma$ and $\sigma \in \Sigma$, we have*

(a) $\beta[\Gamma_1 \cup \Gamma_2] = \beta[\Gamma_1] \cup \beta[\Gamma_2]$;

(b) $\beta[\Gamma_1 \setminus \Gamma_2] = \beta[\Gamma_1] \setminus \beta[\Gamma_2]$;

(c) if $\Gamma_1 \neq \Gamma_2$ then $\beta[\Gamma_1] \neq \beta[\Gamma_2]$, i.e., the image-map $\overline{\beta}$ related to β is injective.

In the case of formulae involving only equality and the Boolean operators of binary union $\cup$ and set difference $\setminus$, the mere injectivity of the map β is enough to preserve satisfiability by partitions. This is proved in the following lemma.

Lemma 2.22. *Let Σ and $\widehat{\Sigma}$ be partitions, V a finite collection of set variables, and $\beta\colon \Sigma \rightarrowtail \widehat{\Sigma}$ an injective map. Further, let $\mathfrak{I}\colon V \to \mathcal{P}(\Sigma)$ be a partition assignment, and ℓ a literal of one of the following types:*

$$x = y \cup z, \qquad x = y \setminus z, \qquad x \neq y, \qquad\qquad (2.8)$$

where x, y, z stand for set variables in V. Then, we have:

(a) $\Sigma/\mathfrak{I} \models \ell \implies \widehat{\Sigma}/\overline{\mathfrak{I}} \models \ell$,
where $\overline{\mathfrak{I}} := \overline{\beta} \circ \mathfrak{I}$ (thus, $\overline{\mathfrak{I}}(x) = \beta[\mathfrak{I}(x)]$, for every $x \in V$).

If, in addition, β is a bijection from Σ onto $\widehat{\Sigma}$, then

(b) $\Sigma/\mathfrak{I} \models \ell \iff \widehat{\Sigma}/\overline{\mathfrak{I}} \models \ell$.

Proof. Let Σ, $\widehat{\Sigma}$, V, β, $\mathfrak{I}$, $\overline{\mathfrak{I}}$, and ℓ be as in the statement of the lemma. We prove (a) by cases.

ℓ is of the type $x = y \cup z$. Then we have:

$$
\begin{array}{llll}
\Sigma/\mathfrak{I} \models x = y \cup z & \implies & \mathfrak{I}(x) = \mathfrak{I}(y) \cup \mathfrak{I}(z) & \text{(by Lemma 2.19(a))} \\
& \implies & \beta[\mathfrak{I}(x)] = \beta[\mathfrak{I}(y) \cup \mathfrak{I}(z)] & \\
& \implies & \beta[\mathfrak{I}(x)] = \beta[\mathfrak{I}(y)] \cup \beta[\mathfrak{I}(z)] & \text{(by Lemma 2.21(a))} \\
& \implies & \overline{\mathfrak{I}}(x) = \overline{\mathfrak{I}}(y) \cup \overline{\mathfrak{I}}(z) & \\
& \implies & \widehat{\Sigma}/\overline{\mathfrak{I}} \models x = y \cup z & \text{(again by Lemma 2.19(a)).}
\end{array}
$$

ℓ is of the type $x = y \setminus z$. In this case we have:

$$
\begin{array}{llll}
\Sigma/\mathfrak{I} \models x = y \setminus z & \implies & \mathfrak{I}(x) = \mathfrak{I}(y) \setminus \mathfrak{I}(z) & \text{(by Lemma 2.19(b))} \\
& \implies & \beta[\mathfrak{I}(x)] = \beta[\mathfrak{I}(y) \setminus \mathfrak{I}(z)] & \\
& \implies & \beta[\mathfrak{I}(x)] = \beta[\mathfrak{I}(y)] \setminus \beta[\mathfrak{I}(z)] & \text{(by Lemma 2.21(b))} \\
& \implies & \overline{\mathfrak{I}}(x) = \overline{\mathfrak{I}}(y) \setminus \overline{\mathfrak{I}}(z) & \\
& \implies & \widehat{\Sigma}/\overline{\mathfrak{I}} \models x = y \setminus z & \text{(again by Lemma 2.19(b)).}
\end{array}
$$

ℓ **is of the type** $x \neq y$. In this last case we have:

$$
\begin{aligned}
\Sigma/\mathfrak{I} \models x \neq y \quad &\Longrightarrow \quad \Sigma/\mathfrak{I} \not\models x = y \cup y \\
&\Longrightarrow \quad \mathfrak{I}(x) \neq \mathfrak{I}(y) && \text{(by Lemma 2.19(a))} \\
&\Longrightarrow \quad \beta[\mathfrak{I}(x)] \neq \beta[\mathfrak{I}(y)] && \text{(by the injectivity of } \beta) \\
&\Longrightarrow \quad \overline{\mathfrak{I}}(x) \neq \overline{\mathfrak{I}}(y) \\
&\Longrightarrow \quad \widehat{\Sigma}/\overline{\mathfrak{I}} \not\models x = y \cup y && \text{(again by Lemma 2.19(a))} \\
&\Longrightarrow \quad \widehat{\Sigma}/\overline{\mathfrak{I}} \models x \neq y.
\end{aligned}
$$

This proves (a).

Let us now assume that the map β is a bijection. We show that (b) follows as well. By part (a), it is enough to show that

$$
\widehat{\Sigma}/\overline{\mathfrak{I}} \models \ell \quad \Longrightarrow \quad \Sigma/\mathfrak{I} \models \ell,
$$

for every literal ℓ of any of the types (2.8).

Thus, assume that $\widehat{\Sigma}/\overline{\mathfrak{I}} \models \ell$. By point (a) of the lemma (but where we use β^{-1} and $\overline{\mathfrak{I}}$ in place of β and $\mathfrak{I}$, respectively), we have

$$
\Sigma/(\overline{\beta^{-1}} \circ \overline{\mathfrak{I}}) \models \ell, \tag{2.9}
$$

where $\overline{\beta^{-1}}$ is the image-map related to β^{-1}. As can be easily verified, $\overline{\beta}$ is a bijection and $\overline{\beta^{-1}} = \overline{\beta}^{-1}$ holds. Hence,

$$
\overline{\beta^{-1}} \circ \overline{\mathfrak{I}} \;=\; \overline{\beta^{-1}} \circ \overline{\beta} \circ \mathfrak{I} \;=\; \overline{\beta}^{-1} \circ \overline{\beta} \circ \mathfrak{I} \;=\; \iota_{_{\mathcal{P}(\Sigma)}} \circ \mathfrak{I} \;=\; \mathfrak{I}
$$

(where, we recall, $\iota_{_{\mathcal{P}(\Sigma)}}$ is the identity function on $\mathcal{P}(\Sigma)$). Thus, $\Sigma/\mathfrak{I} \models \ell$ follows immediately from (2.9). □

2.2.1   Simulations

In order to generalize Lemma 2.22 also to the case of literals of the following types

$$
x \in y, \qquad x \notin y, \qquad x = \mathcal{P}(y), \qquad x = \{y_1, \ldots, y_H\}, \tag{2.10}
$$

we need to impose further constraints on the map β.

Definition 2.23 (Simulation and L-simulation). *A partition $\widehat{\Sigma}$ is said to* SIMULATE *another partition Σ when there is an injection $\beta \colon \Sigma \rightarrowtail \widehat{\Sigma}$, called* SIMULATION, *such that*

$\widehat{\Sigma}$ $\in$-SIMULATES Σ VIA β: $\bigcup \beta[X] \in \bigcup \beta[Y]$ *if and only if* $\bigcup X \in \bigcup Y$, *for* $X, Y \subseteq \Sigma$;

$\widehat{\Sigma}$ $\mathcal{P}$-SIMULATES Σ VIA β: $\bigcup \beta[X] = \mathcal{P}(\bigcup \beta[Y])$ *if* $\bigcup X = \mathcal{P}(\bigcup Y)$, *for* $X, Y \subseteq \Sigma$.

If, in addition, for some $L \geqslant 0$, the injection β satisfies the further constraint

$$
\bigcup X = \{\textstyle\bigcup Y_1, \ldots, \bigcup Y_L\} \quad \Longrightarrow \quad \bigcup \beta[X] = \{\textstyle\bigcup \beta[Y_1], \ldots, \bigcup \beta[Y_L]\}, \tag{2.11}
$$

for all $X, Y_1, \ldots, Y_L \subseteq \Sigma$, then we say that the partition $\widehat{\Sigma}$ L-SIMULATES Σ. In this case, β is called an L-SIMULATION. ∎

(Notice that, when $L = 0$, condition (2.11) is vacuously satisfied.)

The following lemma extends Lemma 2.22 also to literals of the forms (2.10) above.

Lemma 2.24. *Let Σ and $\widehat{\Sigma}$ be partitions, $L \geqslant 0$, and let $\beta \colon \Sigma \rightarrowtail \widehat{\Sigma}$ be an injective map. Further, let $\mathfrak{I} \colon V \to \mathcal{P}(\Sigma)$ be a partition assignment over a given finite collection V of set variables, and let $\overline{\mathfrak{I}} := \overline{\beta} \circ \mathfrak{I}$, where $\overline{\beta}$ is the image-map related to β. Then the following properties hold:*

(a) if β is an $\in$-simulation, then

  (a.1) $\Sigma/\mathfrak{I} \models x \in y \implies \widehat{\Sigma}/\overline{\mathfrak{I}} \models x \in y$, and
  (a.2) $\Sigma/\mathfrak{I} \models x \notin y \implies \widehat{\Sigma}/\overline{\mathfrak{I}} \models x \notin y$,

  for all $x, y \in V$;

(b) if β is a $\mathcal{P}$-simulation, then

$$\Sigma/\mathfrak{I} \models x = \mathcal{P}(y) \implies \widehat{\Sigma}/\overline{\mathfrak{I}} \models x = \mathcal{P}(y),$$

  for all $x, y \in V$; and

(c) if, for all $X, Y_1, \ldots, Y_L \subseteq \Sigma$, the following implication holds

$$\bigcup X = \{\bigcup Y_1, \ldots, \bigcup Y_L\} \implies \bigcup \beta[X] = \{\bigcup \beta[Y_1], \ldots, \bigcup \beta[Y_L]\}, \qquad (2.12)$$

  (namely, condition (2.11) of Definition 2.23 is satisfied), then

$$\Sigma/\mathfrak{I} \models x = \{y_1, \ldots, y_H\} \implies \widehat{\Sigma}/\overline{\mathfrak{I}} \models x = \{y_1, \ldots, y_H\},$$

  for all $x, y_1, \ldots, y_H \in V$, with $0 \leqslant H \leqslant L$.

Proof. Let $\Sigma, \widehat{\Sigma}, \beta, \mathfrak{I}, V$ be as in the hypotheses.

Let us first assume that β is an $\in$-simulation, and let $x, y \in V$. Then we have:

$$
\begin{aligned}
\Sigma/\mathfrak{I} \models x \in y \iff{} & \bigcup \mathfrak{I}(x) \in \bigcup \mathfrak{I}(y) \\
\iff{} & \bigcup \beta[\mathfrak{I}(x)] \in \bigcup \beta[\mathfrak{I}(y)] && \text{(by } \in\text{-simulation)} \\
\iff{} & \bigcup \overline{\mathfrak{I}}(x) \in \bigcup \overline{\mathfrak{I}}(y) \\
\iff{} & \widehat{\Sigma}/\overline{\mathfrak{I}} \models x \in y.
\end{aligned}
$$

Thus, we also have

$$\Sigma/\mathfrak{I} \models x \notin y \iff \Sigma/\mathfrak{I} \not\models x \in y \iff \widehat{\Sigma}/\overline{\mathfrak{I}} \not\models x \in y \iff \widehat{\Sigma}/\overline{\mathfrak{I}} \models x \notin y,$$

proving (a).

Next, let us assume that β is a $\mathcal{P}$-simulation, and let $x, y \in V$. Then we have

$$
\begin{aligned}
\Sigma/\mathfrak{I} \models x = \mathcal{P}(y) \implies{} & \bigcup \mathfrak{I}(x) = \mathcal{P}(\bigcup \mathfrak{I}(y)) \\
\implies{} & \bigcup \beta[\mathfrak{I}(x)] = \mathcal{P}(\bigcup \beta[\mathfrak{I}(y)]) && \text{(by } \mathcal{P}\text{-simulation)} \\
\implies{} & \bigcup \overline{\mathfrak{I}}(x) = \mathcal{P}(\bigcup \overline{\mathfrak{I}}(y)) \\
\implies{} & \widehat{\Sigma}/\overline{\mathfrak{I}} \models x = \mathcal{P}(y),
\end{aligned}
$$

proving (b).

Finally, let us assume that the implication (2.12) holds for all $X, Y_1, \ldots, Y_L \subseteq \Sigma$, and let $x, y_1, \ldots, y_H \in V$, with $0 \leqslant H \leqslant L$. In this case, we have:

$$
\begin{aligned}
\Sigma/\mathfrak{I} \models x = \{y_1, \ldots, y_H\} \quad &\Longrightarrow \quad \bigcup\mathfrak{I}(x) = \{\bigcup\mathfrak{I}(y_1), \ldots, \bigcup\mathfrak{I}(y_H)\} \\
&\Longrightarrow \quad \bigcup\beta[\mathfrak{I}(x)] = \{\bigcup\beta[\mathfrak{I}(y_1)], \ldots, \bigcup\beta[\mathfrak{I}(y_H)]\} \\
&\hspace{6.5cm}\text{(by } L\text{-simulation, as } H \leqslant L) \\
&\Longrightarrow \quad \bigcup\overline{\mathfrak{I}}(x) = \{\bigcup\overline{\mathfrak{I}}(y_1), \ldots, \bigcup\overline{\mathfrak{I}}(y_H)\} \\
&\Longrightarrow \quad \widehat{\Sigma}/\overline{\mathfrak{I}} \models x = \{y_1, \ldots, y_H\},
\end{aligned}
$$

proving (c). This completes the proof of the lemma. $\square$

Lemma 2.24 readily yields that L-simulation preserves the satisfiability of normalized L-bounded MLSSP-conjunctions.

Corollary 2.25. *Let $L \geqslant 0$, and let Σ and $\widehat{\Sigma}$ be partitions such that Σ is L-simulated by $\widehat{\Sigma}$ via an injective map $\beta \colon \Sigma \rightarrowtail \widehat{\Sigma}$. Further, let $\mathfrak{I} \colon V \to \mathcal{P}(\Sigma)$ be a partition assignment over a given finite collection V of set variables. Then, for every normalized L-bounded MLSSP-conjunction Φ such that $\mathrm{Vars}(\Phi) \subseteq V$, we have*

$$
\Sigma/\mathfrak{I} \models \Phi \quad \Longrightarrow \quad \widehat{\Sigma}/\overline{\mathfrak{I}} \models \Phi,
$$

where $\overline{\mathfrak{I}} := \overline{\beta} \circ \mathfrak{I}$ and $\overline{\beta}$ is the image-map related to β.

2.2.2   Imitations

In view of the definition of formative processes and the shadowing relationship among them, we slightly strengthen the notion of simulation. This will simplify the later task to prove that the final partitions of two formative processes, of which one is the shadow of the other, satisfy exactly the same normalized MLSSP-conjunctions.

Definition 2.26 (Imitation). *A partition $\widehat{\Sigma}$ is said to* IMITATE *another partition Σ if there is a bijection $\beta \colon \Sigma \to \widehat{\Sigma}$ such that, for $X \subseteq \Sigma$ and $\sigma \in \Sigma$,*

(i) $\bigcup\beta[X] \in \beta(\sigma)$ *if and only if $\bigcup X \in \sigma$;*

(ii) if $\mathcal{P}^*(\beta[X]) \cap \beta(\sigma) \neq \emptyset$, then $\mathcal{P}^*(X) \cap \sigma \neq \emptyset$;

(iii) if $\mathcal{P}^*(X) \subseteq \bigcup\Sigma$, then $\mathcal{P}^*(\beta[X]) \subseteq \bigcup\beta[\Sigma] = \bigcup\widehat{\Sigma}$.

If, in addition to (i)–(iii), the further condition

(iv) $|\beta(\sigma)| = |\sigma|$, *   when $|\sigma| < \varrho$,*

*is fulfilled, where ϱ is a fixed positive integer, then $\widehat{\Sigma}$ is said to ϱ-*IMITATE Σ.   ■

Remark 2.27. Notice that, when $\widehat{\Sigma}$ and Σ are $\mathcal{P}$-partitions, condition (iii) of Definition 2.26 is implied by condition (i). The reader is asked to prove this fact in Exercise 2.11.
                                                                                                ■

The main technical result of this section shows that imitation implies simulation (and, similarly, ϱ-imitation implies $(\varrho - 1)$-simulation), provided that the imitating partition is transitive.

Lemma 2.28. *Let Σ and $\widehat{\Sigma}$ be partitions, where $\widehat{\Sigma}$ is transitive, and let $\beta\colon \Sigma \to \widehat{\Sigma}$ be a bijection. If $\widehat{\Sigma}$ imitates Σ via β, then $\widehat{\Sigma}$ simulates Σ via β. Furthermore, if $\widehat{\Sigma}$ ϱ-imitates Σ via β, for some $\varrho \geqslant 1$, then $\widehat{\Sigma}$ $(\varrho - 1)$-simulates Σ via β (and, therefore, $\widehat{\Sigma}$ L-simulates Σ via β, for every $0 \leqslant L < \varrho$).*

Proof. Let $\Sigma, \widehat{\Sigma}$ and $\beta\colon \Sigma \to \widehat{\Sigma}$ be as in the hypotheses of the lemma, and assume first that $\widehat{\Sigma}$ imitates Σ via the bijection β. In order to prove that $\widehat{\Sigma}$ simulates Σ via β, we show that $\widehat{\Sigma}$ $\in$-simulates and $\mathcal{P}$-simulates Σ via β.

Concerning $\in$-simulation, let $X, Y \subseteq \Sigma$. Then we have:

$$
\begin{aligned}
\bigcup \beta[X] \in \bigcup \beta[Y] &\iff \bigcup \beta[X] \in \widehat{\sigma}, && \text{for some } \widehat{\sigma} \in \beta[Y] \\
&\iff \bigcup \beta[X] \in \beta(\sigma), && \text{for some } \sigma \in Y \\
&\iff \bigcup X \in \sigma, && \text{for some } \sigma \in Y \\
&\iff \bigcup X \in \bigcup Y,
\end{aligned}
$$

where the third bi-implication follows by condition (i) of Definition 2.26. Hence the partition $\widehat{\Sigma}$ $\in$-simulates Σ via β.

Next, we show that the transitive partition $\widehat{\Sigma}$ $\mathcal{P}$-simulates Σ via β. Thus, let $\bigcup X = \mathcal{P}(\bigcup Y)$, for some $X, Y \subseteq \Sigma$. We first prove that the inclusion

$$\mathcal{P}(\bigcup \beta[Y]) \subseteq \bigcup \beta[X] \tag{2.13}$$

holds. Thus, let $t \in \mathcal{P}(\bigcup \beta[Y])$, hence $t \subseteq \bigcup \beta[Y]$. Let $\widehat{\Sigma}_t \subseteq \beta[Y] \subseteq \widehat{\Sigma}$ be such that $t \in \mathcal{P}^*(\widehat{\Sigma}_t)$. By the bijectivity of β, we have $\beta^{-1}[\widehat{\Sigma}_t] \subseteq Y$. Therefore

$$\mathcal{P}^*(\beta^{-1}[\widehat{\Sigma}_t]) \subseteq \mathcal{P}(\bigcup Y) = \bigcup X \subseteq \bigcup \Sigma, \tag{2.14}$$

so that, by Definition 2.26(iii) and again by the bijectivity of β, $\mathcal{P}^*(\widehat{\Sigma}_t) \subseteq \bigcup \widehat{\Sigma}$ holds. Hence, $t \in \bigcup \widehat{\Sigma}$. Let $\widehat{\sigma}_t$ be the block in $\widehat{\Sigma}$ to which t belongs, and set $\sigma_t := \beta^{-1}(\widehat{\sigma}_t)$. Then, since $\mathcal{P}^*(\widehat{\Sigma}_t) \cap \widehat{\sigma}_t \neq \emptyset$, from Definition 2.26(ii) it follows that $\mathcal{P}^*(\beta^{-1}[\widehat{\Sigma}_t]) \cap \sigma_t \neq \emptyset$. By (2.14), the latter yields $\bigcup X \cap \sigma_t \neq \emptyset$, hence $\sigma_t \in X$. It follows that $t \in \widehat{\sigma}_t \in \beta[X]$, which in turn entails $t \in \bigcup \beta[X]$. This proves (2.13), as t was an arbitrary member of $\mathcal{P}(\bigcup \beta[Y])$.

Next, we prove that the inverse inclusion

$$\bigcup \beta[X] \subseteq \mathcal{P}(\bigcup \beta[Y]) \tag{2.15}$$

holds too. Thus, let now $t \in \bigcup \beta[X]$, and let $\sigma_t \in X$ be such that $t \in \beta(\sigma_t)$. Since $\beta(\sigma_t) \subseteq \bigcup \widehat{\Sigma}$ and $\widehat{\Sigma}$ is transitive, we have $t \subseteq \bigcup \widehat{\Sigma}$. Hence, $t \in \mathcal{P}^*(\beta[\Gamma])$, for some $\Gamma \subseteq \Sigma$. As $\mathcal{P}^*(\beta[\Gamma]) \cap \beta(\sigma_t) \neq \emptyset$, by Definition 2.26(ii) we have $t' \in \mathcal{P}^*(\Gamma) \cap \sigma_t$, for some t'. From $\sigma_t \subseteq \bigcup X = \mathcal{P}(\bigcup Y)$, it follows $t' \in \mathcal{P}^*(Z)$, for some $Z \subseteq Y \subseteq \Sigma$. Since $\{\mathcal{P}^*(T) \mid T \subseteq \Sigma\}$ is a partition, as shown in Lemma 1.29, it follows that $\Gamma = Z$. The latter equality in turn entails $t \subseteq \bigcup \beta[\Gamma] = \bigcup \beta[Z] \subseteq \bigcup \beta[Y]$, so that $t \in \mathcal{P}(\beta[Y])$, proving (2.15), as t was an arbitrary member of $\bigcup \beta[X]$. Finally, (2.13) and (2.15) yield $\bigcup X = \mathcal{P}(\bigcup Y)$, proving that $\widehat{\Sigma}$ $\mathcal{P}$-simulates Σ via β.

Finally we show that if we additionally assume that the partition $\widehat{\Sigma}$ ϱ-imitates Σ via the bijection β, for some $\varrho \geqslant 1$, then $\widehat{\Sigma}$ $(\varrho - 1)$-simulates Σ via β. Having already proved that $\widehat{\Sigma}$ simulates Σ via β, it is enough to show that the constraint (2.11) of Definition 2.23 is fulfilled for $(\varrho - 1)$.

Thus, let $\bigcup X = \{\bigcup Y_1, \ldots, \bigcup Y_H\}$, for $Y_1, \ldots, Y_H \subseteq \Sigma$ (with $H < \varrho$) and $X \subseteq \Sigma$. We need to prove that $\bigcup \beta[X] = \{\bigcup \beta[Y_1], \ldots, \bigcup \beta[Y_H]\}$ holds. Since $\bigcup Y_i \in \bigcup X$ and $\widehat{\Sigma}$ $\in$-simulates Σ via β, then $\bigcup \beta[Y_i] \in \bigcup \beta[X]$, for $i = 1, \ldots, H$, so that

$$\{\textstyle\bigcup \beta[Y_1], \ldots, \bigcup \beta[Y_H]\} \subseteq \bigcup \beta[X]. \tag{2.16}$$

In addition, the injectivity of β yields

$$\left|\{\textstyle\bigcup Y_1, \ldots, \bigcup Y_H\}\right| = \left|\{\textstyle\bigcup \beta[Y_1], \ldots, \bigcup \beta[Y_H]\}\right| \tag{2.17}$$

and

$$\left|\textstyle\bigcup \beta[X]\right| = \sum_{\sigma \in X} |\beta(\sigma)|. \tag{2.18}$$

Finally, since

$$\sum_{\sigma \in X} |\sigma| = \left|\textstyle\bigcup X\right| = \left|\{\textstyle\bigcup Y_1, \ldots, \bigcup Y_H\}\right| \leqslant H < \varrho,$$

for each $\sigma \in X$ we have $|\sigma| < \varrho$, and therefore, by constraint (iv) of Definition 2.26, $|\beta(\sigma)| = |\sigma|$. By (2.18), we have

$$\left|\textstyle\bigcup \beta[X]\right| = \sum_{\sigma \in X} |\beta(\sigma)| = \sum_{\sigma \in X} |\sigma| = \left|\textstyle\bigcup X\right|.$$

By (2.17) and recalling that $\bigcup X = \{\bigcup Y_1, \ldots, \bigcup Y_H\}$, the latter yields

$$\left|\textstyle\bigcup \beta[X]\right| = \left|\{\textstyle\bigcup \beta[Y_1], \ldots, \bigcup \beta[Y_H]\}\right|.$$

Thus, by (2.16), we have

$$\textstyle\bigcup \beta[X] = \{\textstyle\bigcup \beta[Y_1], \ldots, \bigcup \beta[Y_H]\},$$

proving that the transitive partition $\widehat{\Sigma}$ $(\varrho - 1)$-simulates Σ via β. This completes the proof of the lemma. $\square$

From Lemma 2.28 and Corollary 2.25, the following result follows immediately.

Corollary 2.29. *Let Σ and $\widehat{\Sigma}$ be partitions, where $\widehat{\Sigma}$ is transitive, and let $\beta \colon \Sigma \to \widehat{\Sigma}$ be a bijection. Further, let $\mathfrak{I} \colon V \to \mathcal{P}(\Sigma)$ be a partition assignment over a given finite collection V of set variables. If $\widehat{\Sigma}$ ϱ-imitates Σ via β, for some $\varrho \geqslant 1$, then, for every normalized $(\varrho - 1)$-bounded* MLSSP*-conjunction Φ such that* $\mathrm{Vars}(\Phi) \subseteq V$, *we have*

$$\Sigma / \mathfrak{I} \models \Phi \quad \Longrightarrow \quad \widehat{\Sigma} / \overline{\mathfrak{I}} \models \Phi,$$

where $\overline{\mathfrak{I}} := \overline{\beta} \circ \mathfrak{I}$ and $\overline{\beta}$ is the image-map related to β.

The preceding result can be extended to MLSSPF-conjunctions as follows.

Lemma 2.30. *Let Σ and $\widehat{\Sigma}$ be partitions, where $\widehat{\Sigma}$ is transitive, and let V be a finite collection of set variables. Assume that the following conditions are fulfilled, for a given bijection $\beta \colon \Sigma \rightarrowtail \widehat{\Sigma}$:*

(a) $\widehat{\Sigma}$ ϱ-imitates Σ via β, for some $\varrho \geqslant 1$; and

(b) $|\sigma| < \aleph_0$ if and only if $|\beta(\sigma)| < \aleph_0$, for every $\sigma \in \Sigma$.

Then, for every normalized $(\varrho - 1)$-bounded MLSSPF*-conjunction Φ such that* $\mathrm{Vars}(\Phi) \subseteq V$ *and every partition assignment $\mathfrak{I}\colon V \to \mathcal{P}(\Sigma)$, we have*

$$\Sigma/\mathfrak{I} \models \Phi \quad \Longrightarrow \quad \widehat{\Sigma}/\overline{\mathfrak{I}} \models \Phi,$$

where $\overline{\mathfrak{I}} := \overline{\beta} \circ \mathfrak{I}$ and $\overline{\beta}$ is the image-map related to β.

(The proof of the lemma is left to the reader as an exercise.)

Most of the time, we shall work with $\mathcal{P}$-partitions. In the case of $\mathcal{P}$-partitions, Lemma 2.28 yields:

Corollary 2.31. *Let $\Sigma, \widehat{\Sigma}$ be $\mathcal{P}$-partitions. If $\widehat{\Sigma}$ imitates Σ, then $\widehat{\Sigma}$ simulates Σ. If, furthermore, $\widehat{\Sigma}$ ϱ-imitates Σ, for some $\varrho \geqslant 1$, then $\widehat{\Sigma}$ $(\varrho - 1)$-simulates Σ.*

In addition, a result similar to Corollary 2.29 holds when Σ and $\widehat{\Sigma}$ are $\mathcal{P}$-partitions.

Remark 2.32. In the particular case in which a $\mathcal{P}$-partition $\widehat{\Sigma}$ $\in$-simulates another $\mathcal{P}$-partition Σ via a *bijection* β, the map β sends the special block of Σ into the special block of $\widehat{\Sigma}$. Indeed, let σ^* and $\widehat{\sigma}^*$ be the special blocks of Σ and $\widehat{\Sigma}$, respectively. By Lemma 1.24(c), we have $\bigcup(\Sigma \setminus \{\sigma^*\}) \in \sigma^*$. Hence, the property of $\in$-simulation yields $\bigcup \beta[\Sigma \setminus \{\sigma^*\}] \in \beta(\sigma^*)$. By the bijectivity of β and since Σ and $\widehat{\Sigma}$ are partitions, we have

$$\bigcup \beta[\Sigma \setminus \{\sigma^*\}] = \bigcup \left(\beta[\Sigma] \setminus \{\beta(\sigma^*)\} \right) = \bigcup \beta[\Sigma] \setminus \beta(\sigma^*) = \bigcup \widehat{\Sigma} \setminus \beta(\sigma^*),$$

hence $\bigcup \widehat{\Sigma} \setminus \beta(\sigma^*) \in \beta(\sigma^*)$ holds. Thus, again by Lemma 1.24(c), it follows that $\beta(\sigma^*)$ is the special block of $\widehat{\Sigma}$, i.e., $\beta(\sigma^*) = \widehat{\sigma}^*$.   ∎

2.3   A Simple Case Study: The Theory BST

As a first step towards the far more challenging decision problems for MLSSP and MLSSPF, which will be considered in detail in the second part of the book, here we illustrate the very simple case of the decidability of the very simple theory BST (for Boolean Set Theory). The formulae of BST are the propositional combinations of atoms of the following four types:

$$x = y \cup z, \qquad x = y \cap z, \qquad x = y \setminus z, \qquad x \subseteq y,$$

where x, y, z stand for set variables or the constant $\varnothing$. Thus, BST is the subtheory of MLS obtained by dropping the membership literals.

By way of a reduction of the type presented in Section 2.1.2, for decidability purposes we can limit ourselves to conjunctions of BST-literals of the following types

$$x = y \cup z, \qquad x = y \setminus z, \qquad x \neq y \qquad\qquad (2.19)$$

(here x, y, z stand just for *set variables*), which we call NORMALIZED BST-CONJUNCTIONS.

Since every normalized BST-conjunction Φ containing no negative literal of type $x \neq y$ is plainly satisfiable by the *null* set assignment (namely, the set assignment $M_\varnothing$ such that $M_\varnothing x := \emptyset$, for every variable v occurring in Φ), in the rest of the section, without loss of generality, we assume that all normalized BST-conjunctions we shall consider contain at least one negative literal of type $x \neq y$. It is a general fact that:

Fact 1:   Every $\mathcal{S}$-formula Ψ involving exactly n distinct set variables is satisfiable if and only if it is satisfied by some partition with $(2^n - 1)$ blocks.

In addition, in the restricted case of normalized BST-conjunction (and, in fact, for any BST-formula), Lemma 2.22 yields:

Fact 2:   A normalized BST-conjunction Φ is satisfied by a partition Σ if and only if it is satisfied by any partition $\widehat{\Sigma}$ such that $\left|\widehat{\Sigma}\right| \geqslant |\Sigma|$.

Bt Facts 1 and 2, we have:

A normalized BST-conjunction Φ with n distinct set variables is satisfiable if and only if it is satisfied by any partition with $(2^n - 1)$ blocks (regardless of the internal structure of the blocks).

This is evidently sufficient to establish the decidability of BST, and justifies the following satisfiability test for normalized BST-conjunctions:

```
1: procedure BST-SatTest(Φ);           ▷ Φ is a normalized BST-conjunction
2:     - let n be the number of distinct set variables occurring in Φ;
3:     - let Σ be any partition of cardinality 2ⁿ − 1;
4:     if Σ satisfies Φ then
5:         return "Φ is satisfiable";
6:     else
7:         return "Φ is unsatisfiable";
8:     end if;
9: end procedure;
```

The partition Σ picked at step 3 could be, for instance, $\left\{ \emptyset^1, \emptyset^2, \ldots, \emptyset^{(2^n-1)} \right\}$, where we are using the nesting notation $\emptyset^h$ introduced in Example 1.12. Then the effectiveness of procedure BST-SatTest follows, since to check at step 4 whether Σ satisfies Φ is a task that can be accomplished in a finite number of steps.

Summarizing, we obtain:

Corollary 2.33.   *The theory BST has a solvable satisfiability problem.*

2.3.1   Complexity Issues

For a BST-conjunction Φ with n distinct set variables, the complexity of the decision procedure BST-SatTest is $\Omega\left(2^{n \cdot 2^n - n}\right)$, since $2^{n \cdot 2^n - n}$ is the number of different maps $\mathfrak{I}$ from the variables of Φ into $\mathcal{P}(\{\, \emptyset^1, \emptyset^2, \ldots, \emptyset^{(2^n-1)} \,\})$. This in turn yields a nondeterministic exponential-time algorithm. However, a more accurate analysis allows one to show that the satisfiability problem for BST admits a nondeterministic polynomial-time decision procedure: hence, it falls into the complexity class NP.

Remark 2.34. We recall that, roughly speaking, NP is the class of all decision problems whose positive answers can be certified in deterministic polynomial time. Some of the decision problems in NP are harder than the others, in the sense that any deterministic polynomial-time solution for them would yield a deterministic polynomial-time solution to *all* the decision problems in NP. These are the so-called NP-COMPLETE PROBLEMS. Among them, we mention the PROPOSITIONAL SATISFIABILITY PROBLEM (SAT), namely, the problem of establishing, for any given propositional formula, whether or not it is satisfied by some propositional valuation. The decision problem SAT is the progenitor of a large collection of NP-complete problems. Particularly useful to our analysis, will be the NP-complete problem 3-SAT, which is a restriction of the problem SAT to propositional formulae in conjunctive normal form where each conjunct contains *exactly* three distinct literals, namely, propositional variables or their negations.

The typical pattern for proving that a given decision problem D is NP-complete consists in the following two steps:

(I) prove that the problem D belongs to the class NP;

(II) prove that the problem D is at least as difficult as any other problem in NP, namely, that it is NP-HARD.

Step (I) can be performed by showing that each positive instance in D admits a polynomial-time certification. Concerning step (II), one can select any known NP-complete problem D_{NP} and reduce it polynomially to D in such a way that any deterministic polynomial-time solution to D would yield a deterministic polynomial-time solution to D_{NP} (*reduction technique*).

Plainly, every decision problem solvable in deterministic polynomial time belongs to the class NP. The converse is not known, and it is one of the most important and intriguing open problems in theoretical computer science, known as the P = NP question. According to current knowledge, any deterministic implementation of a nondeterministic polynomial-time algorithm for an NP-complete problem runs in (at least) exponential time. Thus, by proving that a given decision problem D is NP-complete it practically means that it is very unlikely that a polynomial-time algorithm for D can ever be found. The interested reader may refer to a number of sources (see, for instance, [GJ90, Sip12]). ∎

In the case of the satisfiability problem for normalized BST-conjunctions Φ, verification certificates will have the form of 'small' models of Φ induced by the partition assignment $\mathfrak{I}\colon \mathrm{Vars}(\Phi) \to \mathcal{P}(\overline{\Sigma})$, where $\overline{\Sigma}$ is a partition of size $|\mathrm{Vars}(\Phi)| - 1$. The correct partition assignment $\mathfrak{I}$ can thus be guessed in nondeterministic polynomial time, yielding an overall nondeterministic polynomial-time decision procedure for normalized BST-conjunctions.

Our plan is then to show that any normalized BST-conjunction involving m distinct set variables is satisfiable if and only if it is satisfied by a partition $\overline{\Sigma}$ of size at most $(m-1)$.

Preliminary to that, we first state the following technical result, whose proof is left to the reader as Exercise 2.13.

Lemma 2.35. *Let Σ and V be, respectively, a finite partition and a finite collection of set variables, and let $\mathfrak{I}\colon V \to \mathcal{P}(\Sigma)$ be a partition assignment. Further, assume that $\Sigma^* \subseteq \Sigma$, and let $\mathfrak{I}^*\colon V \to \mathcal{P}(\Sigma)$ be the map defined as follows, for $v \in V$:*

$$\mathfrak{I}^*(v) := \mathfrak{I}(v) \cap \Sigma^* \,.$$

Then, for every positive BST-*literal ℓ of type*

$$x = y \cup z \qquad or \qquad x = y \setminus z$$

whose variables belong to V, we have

$$\Sigma/\mathfrak{I} \models \ell \quad \Longrightarrow \quad \Sigma^*/\mathfrak{I}^* \models \ell.$$ ■

The following lemma readily implies the small model property sought for for normalized BST-conjunctions.

Lemma 2.36. *Any normalized* BST-*conjunction Φ with n distinct set variables in its negative part, namely, the conjunction of the negative literals of type $x \neq y$ in Φ, is satisfiable if and only if it is satisfied by a partition $\overline{\Sigma}$ of size at most $(n-1)$.*

Proof. Plainly, we just need to prove the sufficiency part of the lemma. Thus, let Φ be a satisfiable normalized BST-conjunction, and assume that its negative part Φ^- involves $n \geqslant 1$ distinct set variables. Further, let Σ be any partition such that $\Sigma/\mathfrak{I} \models \Phi$, for some partition assignment $\mathfrak{I} : \mathrm{Vars}(\Phi) \to \mathcal{P}(\Sigma)$. Following a similar approach to that of [CF95] (cf. also [PPR97]), we say that a subset $\Gamma \subseteq \Sigma$ WEAKLY DISTINGUISHES a set of variables V with respect to Φ if $\mathfrak{I}(x) \cap \Gamma \neq \mathfrak{I}(y) \cap \Gamma$, for every pair of variables $x, y \in V$ such that $x \neq y$ is in Φ.[10]

Let $\overline{\Gamma}$ be the collection of blocks in Σ returned by the following procedure:

```
 1: procedure WEAKLY-DISTINGUISH(Φ, Σ, ℑ);
 2:     Γ ← ∅;  Done ← ∅;
 3:     - let Φ⁻ be the conjunction of the literals of type x ≠ y in Φ;
 4:     for v ∈ Vars(Φ⁻) do
 5:         if Γ does not weakly distinguish Done ∪ {v} then
 6:             - let v' ∈ Done such that v ≠ v' is in Φ but ℑ(v) ∩ Γ = ℑ(v') ∩ Γ;
 7:             - pick σ ∈ Σ such that σ ∈ ℑ(v) ⟺ σ ∉ ℑ(v');
 8:             Γ ← Γ ∪ {σ};
 9:             Done ← Done ∪ {v};
10:         end if;
11:     end for;
12:     return Γ;
13: end procedure;
```

Then $\overline{\Gamma}$ weakly distinguishes the variables in Φ, and $|\overline{\Gamma}| \leqslant n - 1$ holds. Indeed, the **for-loop** 4–11 is executed n times, and at each iteration at most one block is added to Γ (at line 8). In addition, Γ is initialized to the empty set (at line 2) and no element is added to it during the first iteration. Hence, $|\overline{\Gamma}| \leqslant n - 1$ holds. To prove that $\overline{\Gamma}$ weakly distinguishes the set of variables $\mathrm{Vars}(\Phi)$, we preliminarily observe that if a set of blocks Γ distinguishes a set of variables V, then so does each of its supersets. Thus, assume inductively that the set Γ (in procedure WEAKLY-DISTINGUISH) weakly distinguishes the variables in Done. Then, if at line 5 the set Γ does not weakly distinguish Done $\cup \{v\}$ (w.r.t. Φ), there must be exactly one $v' \in$ Done such that $\mathfrak{I}(v) \neq \mathfrak{I}(v')$ but $\mathfrak{I}(v) \cap \Gamma = \mathfrak{I}(v') \cap \Gamma$. It follows that, upon extending Γ with some block $\sigma \in \Sigma$ such that $\sigma \in \mathfrak{I}(v) \iff \sigma \notin \mathfrak{I}(v')$, the set $\Gamma \cup \{\sigma\}$

---

[10]A subset $\Gamma \subseteq \Sigma$ DISTINGUISHES a set of variables $V \subseteq \mathrm{Vars}(\Phi)$ if $\mathfrak{I}(x) \cap \Gamma \neq \mathfrak{I}(y) \cap \Gamma$, for every pair of variables $x, y \in V$.

weakly distinguishes all the pairs in $\mathsf{Done} \cup \{v\}$ involving v' (w.r.t. Φ), and therefore all the pairs in $\mathsf{Done} \cup \{v\}$.

Let us now consider the partition assignment $\mathfrak{J}^* \colon \mathsf{Vars}(\Phi) \to \mathcal{P}(\overline{\Gamma})$ defined by

$$\mathfrak{J}^*(v) := \mathfrak{J}(v) \cap \overline{\Gamma}, \tag{2.20}$$

for $v \in \mathsf{Vars}(\Phi)$. By Lemma 2.35, Γ satisfies all positive literals in Φ via the partition assignment $\mathfrak{J}^*$. In addition, recalling that Σ satisfies Φ via $\mathfrak{J}$, then, for each negative literal $x \neq y$ in Φ, we have $\mathfrak{J}(x) \neq \mathfrak{J}(y)$, by Lemma 2.19. The latter implies $\mathfrak{J}^*(x) \neq \mathfrak{J}^*(y)$, since $\overline{\Gamma}$ weakly distinguishes the variables in $\mathsf{Vars}(\Phi)$ (w.r.t. Φ), so that, again by Lemma 2.19, $\overline{\Gamma}$ satisfies $x \neq y$ via the map $\mathfrak{J}^*$. Thus, $\overline{\Gamma}$ satisfies all the negative literals in Φ via the map $\mathfrak{J}^*$. □

The above discussion can be summarized as follows.

Lemma 2.37. *The satisfiability problem for normalized* BST*-conjunctions belongs to the class* NP*, namely, it can be solved in nondeterministic polynomial time.*

Remark 2.38. Notice that if a satisfiable normalized BST-conjunction Φ contains exactly one negative literal of type $x \neq y$, then the procedure WEAKLY-DISTINGUISH returns a distinguishing set $\overline{\Gamma}$ of cardinality 1, thus such that $\overline{\Gamma} = \{\overline{\sigma}\}$, for some $\overline{\sigma} \in \Sigma$. It follows that the map $\mathfrak{J}^*$ defined by (2.20) induces a set assignment $M_{\mathfrak{J}^*}$ over $\mathsf{Vars}(\Phi)$ that satisfies Φ and is such that either $M_{\mathfrak{J}^*} v = \emptyset$ or $M_{\mathfrak{J}^*} v = \overline{\sigma}$, for every $v \in \mathsf{Vars}(\Phi)$. ∎

Next we show that the satisfiability problem for normalized BST-conjunctions is NP-hard, by reducing to it the problem 3-SAT.[11] An example of a satisfiable 3-SAT instance is

$$(\mathsf{P}_1 \vee \neg\mathsf{P}_2 \vee \mathsf{P}_3) \wedge (\neg\mathsf{P}_1 \vee \neg\mathsf{P}_2 \vee \mathsf{P}_3) \wedge (\mathsf{P}_1 \vee \mathsf{P}_2 \vee \mathsf{P}_3), \tag{2.21}$$

where $\mathsf{P}_1, \mathsf{P}_2, \mathsf{P}_3$ are propositional variables. Indeed, a propositional valuation[12] $\mathfrak{v}$ which satisfies (2.21) is the one such that $\mathfrak{v}(\mathsf{P}_1) = \mathbf{t}, \mathfrak{v}(\mathsf{P}_2) = \mathfrak{v}(\mathsf{P}_3) = \mathbf{f}$, since

$$\mathfrak{v}(\mathsf{P}_1 \vee \neg\mathsf{P}_2 \vee \mathsf{P}_3) = \mathfrak{v}(\neg\mathsf{P}_1 \vee \neg\mathsf{P}_2 \vee \mathsf{P}_3) = \mathfrak{v}(\mathsf{P}_1 \vee \mathsf{P}_2 \vee \mathsf{P}_3) = \mathbf{t}\,.$$

In order to reduce the 3-SAT problem to the satisfiability problem for normalized BST-conjunctions, it is enough to exhibit a polynomial-time algorithm that, for any given 3-SAT instance F, constructs a normalized BST-conjunction Φ_F such that:

F is propositionally satisfiable   if and only if Φ_F is satisfiable by a set model.

Thus, let

$$F \overset{\text{Def}}{:=} C_1 \wedge \ldots \wedge C_m \tag{2.22}$$

be a 3-SAT instance, where $C_i \overset{\text{Def}}{:=} L_{i1} \vee L_{i2} \vee L_{i3}$ and the L_{ij}'s are propositional literals. Let $\mathsf{P}_1, \ldots, \mathsf{P}_n$ be the distinct propositional variables in F, and let $X_1, \ldots, X_n, X_U$ be $(n+1)$

---

[11] A similar approach appears in [AP88] for the set containment inference problem.
[12] We recall that a propositional valuation is an assignment of truth values to propositional variables.

distinct set variables. For $i = 1, \ldots, m$ and $j = 1, 2, 3$, we put

$$T_{ij} := \begin{cases} X_k & \text{if } L_{ij} = \mathsf{P}_k, \text{ for some } k \\ X_U \setminus X_k & \text{if } L_{ij} = \neg \mathsf{P}_k, \text{ for some } k, \end{cases}$$

and then put

$$\mathcal{C}_i \overset{\text{Def}}{:=} (X_U = T_{i1} \cup T_{i2} \cup T_{i3}).$$

Finally, we put

$$\Phi_F^* \overset{\text{Def}}{:=} \bigwedge_{i=1}^{m} \mathcal{C}_i \wedge \bigwedge_{j=1}^{n} X_j \subseteq X_U \wedge X_U \neq \varnothing. \tag{2.23}$$

It is easy to show that if F is propositionally satisfiable, then the BST-conjunction Φ_F^* is satisfiable by a set model. In addition, the above construction can be carried out in linear time in $|F|$. Conversely, if Φ_F^* is satisfiable, then, by Remark 2.38, it is satisfied by a set assignment M such that $Mv \in \{\emptyset, \overline{\sigma}\}$, for some nonempty set $\overline{\sigma}$ and for every $v \in \text{Vars}(\Phi_F^*)$, since Φ_F^* has only one negative literal. Thus, if $\mathfrak{v}$ is the propositional valuation defined by

$$\mathfrak{v}(\mathsf{P}_i) := \begin{cases} \mathbf{t} & \text{if } MX_i \neq \emptyset \\ \mathbf{f} & \text{otherwise,} \end{cases}$$

for $i = 1, \ldots, n$, then it easily follows that $\mathfrak{v}$ satisfies the 3-SAT instance (2.22).

Notice that the BST-conjunction Φ_F^* can be easily converted in linear time into an equisatisfiable normalized BST-conjunction Φ_F. This completes the proof that the NP-complete problem 3-SAT can be reduced in linear time to the satisfiability problem for normalized BST-conjunctions.

In view of Lemma 2.37, the above considerations prove the following result:

Theorem 2.39. *The satisfiability problem for normalized* BST-*conjunctions is* NP-*complete.*

2.4   Expressiveness Results

In this section we prove, for some theories, that certain operations and relations, whose related interpreted operators and relators are not among their primitive symbols, are expressible in them anyway.[13] In such cases, we shall say indifferently that the operations and relations under consideration, or their related interpreted operators and relators, are expressible in the given theory. More specifically, we show the following expressiveness results:

- the singleton operator is expressible in MLSC and in MLSuC;

- hereditarily finite sets are expressible in MLSC, MLSuC, and MLSP (for the former two theories, this follows immediately from the preceding result);

---

[13]We make the following distinction between *operations* and *operators*: an n-ary operation over $\mathcal{V}$ is a (class) map whose input is an n-tuple of sets and whose output is a set; instead, an n-ary operator is an n-place function symbol that needs to be interpreted as a specific n-ary operation. We also make a similar distinction between *relations* and *relators*: an n-ary relation over $\mathcal{V}$ is a class of n-tuples, whereas an n-ary relator is an n-place predicate symbol that needs to be interpreted as a specific n-ary relation.

- terms of the form $\mathcal{P}^*(\{x_1, \ldots, x_k\})$ are expressible in MLSP;

- positive instances of the cardinality comparison relation $|X| = |Y|$, asserting that X and Y have the same cardinality, are expressible in the extension $\mathsf{MLSuC^+}$ of MLSuC with literals of the form $x = \uplus y$;

- positive instances of the finiteness unary relation $Finite(X)$, asserting that X is finite, are expressible in the extension $\mathsf{MLSuC^+}$ of MLSuC.

The latter two results will allow us to prove that the satisfiability problem for the theory $\mathsf{MLSuC^+}$ is undecidable. Instead, the first expressiveness result implies that MLSuC is not *rank dichotomic*, in the sense that, as we shall see below, there are normalized conjunctions of MLSuC that admit models of arbitrarily large finite rank, whereas they do not admit any model of infinite rank.

For the sake of accuracy, we define in precise terms what we mean by saying that an operation or a relation over sets is expressible in a fragment of $\mathcal{S}$.

Definition 2.40 (Expressiveness of operations). *Let $\mathfrak{F}$ be a subtheory of $\mathcal{S}$, and $\mathbf{P}$ a given n-ary operation over the universe $\mathcal{V}$ of all sets. We say that $\mathbf{P}$ is* EXPRESSIBLE IN $\mathfrak{F}$ *(via the variables $x_1, \ldots, x_n, y$) when there exists an $\mathfrak{F}$-formula $\psi_{\mathbf{p}}(x_1, \ldots, x_n, y)$, possibly involving other (auxiliary) variables besides $x_1, \ldots, x_n, y$, such that the following two conditions hold:*

(a) for every set assignment M satisfying $\psi_{\mathbf{p}}(x_1, \ldots, x_n, y)$, the equality

$$My = \mathbf{P}(Mx_1, \ldots, Mx_n)$$

holds, and

(b) for all sets $X_1, \ldots, X_n$, there exists a set assignment $M_{\vec{X}}$ over the variables of $\psi_{\mathbf{p}}$ such that:

  - $M_{\vec{X}}(y) = \mathbf{P}(X_1, \ldots, X_n)$,
  - $M_{\vec{X}}(x_i) = X_i$, *for $i = 1, \ldots, n$, and*
  - $M_{\vec{X}} \models \psi_{\mathbf{p}}(x_1, \ldots, x_n, y)$.

In such a case, we say that the $\mathfrak{F}$-formula $\psi_{\mathbf{p}}(x_1, \ldots, x_n, y)$ EXPRESSES $\mathbf{P}$. ∎

A similar definition holds also for relations.

Definition 2.41 (Expressiveness of relations). *Let $\mathfrak{F}$ be a subtheory of $\mathcal{S}$, and $\mathbf{R}$ a given n-ary relation over the universe $\mathcal{V}$ of all sets. We say that $\mathbf{R}$ is* EXPRESSIBLE IN $\mathfrak{F}$ *(via the variables $x_1, \ldots, x_n$) when there exists an $\mathfrak{F}$-formula $\psi_{\mathbf{R}}(x_1, \ldots, x_n)$, possibly involving other (auxiliary) variables besides $x_1, \ldots, x_n$, such that the following two conditions hold:*

(a) for every set assignment M satisfying $\psi_{\mathbf{R}}(x_1, \ldots, x_n)$, the equality

$$\mathbf{R}(Mx_1, \ldots, Mx_n)$$

holds, and

(b) for all sets $X_1, \ldots, X_n$, there exists a set assignment $M_{\vec{X}}$ over the variables of $\psi_{\mathbf{R}}$ such that:

  - $\mathbf{R}(X_1, \ldots, X_n)$ holds,
  - $M_{\vec{X}}(x_i) = X_i$, for $i = 1, \ldots, n$, and
  - $M_{\vec{X}} \models \psi_{\mathbf{R}}(x_1, \ldots, x_n)$.

In such a case, we say that the $\mathfrak{F}$-formula $\psi_{\mathbf{R}}(x_1, \ldots, x_n)$ EXPRESSES $\mathbf{R}$.   ∎

2.4.1   Expressing the Singleton Operator

We show that the singleton operator is expressible in both MLSC and MLSuC (see [CCP89]). As a consequence, it will follow that also hereditarily finite sets are expressible both in MLSC and in MLSuC.

We start with MLSC. Let $\psi_{\times}(x, y)$ be the conjunction of the following MLSC-literals

$$\begin{array}{llll}
(\ell_1) \quad z = y' \times y' & (\ell_2) \quad w \in z & (\ell_3) \quad y \in w & (\ell_4) \quad y' \in w \\
(\ell_5) \quad x \in y & (\ell_6) \quad x \in y' & (\ell_7) \quad x' \in y' & (\ell_8) \quad x' \notin y,
\end{array} \qquad (2.24)$$

and let M be any set assignment satisfying $\psi_{\times}(x, y)$. We claim that $My = \{Mx\}$ must hold.

By literals (ℓ_1) and (ℓ_2), Mw must be an ordered pair. Thus, either Mw is the singleton of a singleton, or it contains two elements, namely, an unordered pair and a singleton subset of it. By literals (ℓ_3) and (ℓ_4), we have $My, My' \in Mw$. Since, literals (ℓ_7) and (ℓ_8) yield that $My \neq My'$, it follows that $Mw = \{My, My'\}$. Finally, literals (ℓ_5)–(ℓ_8) entail that $Mx \neq Mx'$, $Mx, Mx' \in My'$, and $Mx \in My$. Thus, $My = \{Mx\}$, as claimed (and $My' = \{Mx, Mx'\}$).

The converse, namely that, for every set X, there exists a set assignment M_X over the variables of $\psi_{\times}(x, y)$ that satisfies $\psi_{\times}(x, y) \wedge y = \{x\}$ and is such that $Mx = My$ and $My = \{X\}$ hold, is left to the reader as Exercise 2.19.

Thus, the singleton operator is expressible in MLSC.

To show that the singleton operation is expressible in MLSuC too, consider the conjunction $\psi_{\otimes}(x, y)$ of the following MLSuC-literals:

$$\begin{array}{lll}
(\ell'_1) \quad z = y' \otimes y', & (\ell'_2) \quad y' \in z, & (\ell'_3) \quad y \in y', \\
(\ell'_4) \quad y \subseteq y', & (\ell'_5) \quad x \in y. &
\end{array} \qquad (2.25)$$

It is an easy matter to prove that the singleton is expressed by the MLSuC-conjunction $\psi_{\otimes}(x, y)$ (see Exercise 2.20).

Summarizing, we have:

Lemma 2.42. *The singleton operator is expressible both in* MLSC *and in* MLSuC.

2.4.2   Expressing Hereditarily Finite Sets

We show now that hereditarily finite sets are expressible in MLSC, MLSuC, and MLSP. In the case of the theories MLSC and MLSuC, this follows immediately from Lemma 2.42, as singletons are expressible in MLSC and MLSuC; hence, starting from the empty set and using the representing formula $\psi_\times(x,y)$ or $\psi_\otimes(x,y)$ in the appropriate manner, it is possible to express every hereditarily finite set.

Example 2.43. Let us show, for instance, that the hereditarily finite set $\mathfrak{h} := \{\emptyset, \{\emptyset, \{\emptyset\}\}\}$ is expressible in MLSC. Consider the conjunction Φ of the following MLSS-literals:

$$y_1 = \{\varnothing\}, \quad y_2 = \{y_1\}, \quad y_3 = y_1 \cup y_2, \quad y_4 = \{y_3\}, \quad y = y_2 \cup y_4. \tag{2.26}$$

It is easy to see that Φ is satisfiable, and that $My = \mathfrak{h}$ holds, for every model M of Φ. Let $\Phi_\times$ be the conjunction of the following MLSC-formulae:

$$\psi_\times(\varnothing, y_1), \quad \psi'_\times(y_1, y_2), \quad y_3 = y_1 \cup y_2, \quad \psi''_\times(y_3, y_4), \quad y = y_2 \cup y_4, \tag{2.27}$$

where $\psi_\times$ is the conjunction of the literals (2.24), and $\psi'_\times$ and $\psi''_\times$ are variants of $\psi_\times$ obtained by renaming its auxiliary variables in such a way that $\psi_\times$, $\psi'_\times$, and $\psi''_\times$ do not share any auxiliary variable. Plainly, the literals in (2.27) are obtained from the literals (2.26) by expressing each literal $t_1 = \{t_2\}$ in terms of the MLSC-formula $\psi_\times$. Thus, the MLSC-formula $\Phi_\times$ expresses the hereditarily finite set $\mathfrak{h}$, in the sense that $\Phi_\times$ is satisfiable, and $My = \mathfrak{h}$ holds, for every model M of $\Phi_\times$.

Likewise, the conjunction $\Phi_\otimes$ of the following MLSuC-formulae

$$\psi_\otimes(\varnothing, y_1), \quad \psi'_\otimes(y_1, y_2), \quad y_3 = y_1 \cup y_2, \quad \psi''_\otimes(y_3, y_4), \quad y = y_2 \cup y_4$$

(where $\psi_\otimes$ is the conjunction of the literals (2.25) and, much as above, $\psi'_\otimes$ and $\psi''_\otimes$ are variants of $\psi_\otimes$ obtained by renaming its auxiliary variables in such a way that $\psi_\otimes$, $\psi'_\otimes$, and $\psi''_\otimes$ do not share any auxiliary variable) expresses the hereditarily finite set $\mathfrak{h}$. ∎

Next, we show now that hereditarily finite sets are expressible in MLSP. For every set s, we plainly have

$$s = \bigcup_{s' \in s} \{s'\},$$

so that

$$\mathcal{P}(s) = \bigcup_{s'' \subseteq s} \{s''\}$$

holds, where the singletons on the right-and-side are all pairwise disjoint. Thus, we have

$$\{s\} = \mathcal{P}(s) \setminus \bigcup_{s'' \subsetneq s} \{s''\}$$
$$= \mathcal{P}\Big(\bigcup_{s' \in s} \{s'\}\Big) \setminus \bigcup_{s'' \subsetneq s} \{s''\}.$$

The latter equation implies that the singleton of any finite set s can be expressed by a formula involving the singletons of its elements and of its proper subsets, and the operations of binary union, set difference, and powerset. This is a key observation towards a recursive representation of hereditarily finite sets with MLSP-conjunctions of positive literals of the following types:

$$x = y \cup z, \quad x = y \setminus z, \quad x = \mathcal{P}(y). \tag{2.28}$$

In our representation, we shall initially use only variables indexed by singletons of hereditarily finite sets, namely, of the form $x_{\{\mathfrak{h}\}}$: other auxiliary variables, indexed by hereditarily finite sets, will enter our representing formulae $\varphi_{\{\mathfrak{h}\}}$ during the phase of elimination of compound terms. It will turn out that each representing formula $\varphi_{\{\mathfrak{h}\}}$, with $\mathfrak{h} \in \mathsf{HF}$, is satisfiable and enjoys the following *faithfulness condition*:[14]

(FC) for all hereditarily finite sets $\mathfrak{h}$ and $\mathfrak{h}'$ such that $x_{\{\mathfrak{h}'\}}$ occurs in $\varphi_{\{\mathfrak{h}\}}$,[15] we have

$$\models \varphi_{\{\mathfrak{h}\}} \;\rightarrow\; x_{\{\mathfrak{h}'\}} = \{\mathfrak{h}'\}\,.$$

For each $\mathfrak{h} \in \mathsf{HF}$, we recursively define the representing MLSP-formula $\varphi_{\{\mathfrak{h}\}}$ of $\mathfrak{h}$ by putting:

$$\varphi_{\{\mathfrak{h}\}} \stackrel{\text{Def}}{:=} \begin{cases} \left(x_{\{\emptyset\}} = \mathcal{P}\!\left(x_{\{\emptyset\}} \setminus x_{\{\emptyset\}}\right)\right) & \text{if } \mathfrak{h} = \emptyset \\[2mm] \left(x_{\{\mathfrak{h}\}} = \mathcal{P}\!\left(\bigcup_{\mathfrak{h}'\in\mathfrak{h}} x_{\{\mathfrak{h}'\}}\right) \setminus \bigcup_{\mathfrak{h}''\subsetneq\mathfrak{h}} x_{\{\mathfrak{h}''\}}\right) \wedge \bigwedge_{\mathfrak{h}'\in\mathfrak{h}} \varphi_{\{\mathfrak{h}'\}} \wedge \bigwedge_{\mathfrak{h}''\subsetneq\mathfrak{h}} \varphi_{\{\mathfrak{h}''\}} & \text{if } \mathfrak{h} \neq \emptyset\,. \end{cases}$$

$$(2.29)$$

To prove that the recursive definition (2.29) is well-given, it is enough to exhibit a well-ordering $\prec$ of HF such that the following properties hold, for $\mathfrak{h} \in \mathsf{HF}$:

(P1) $\mathfrak{h}' \prec \mathfrak{h}$, for every $\mathfrak{h}' \in \mathfrak{h}$, and

(P2) $\mathfrak{h}'' \prec \mathfrak{h}$, for every $\mathfrak{h}'' \subsetneq \mathfrak{h}$.

Notice that any total ordering $\prec$ of HF such that

(i) $\prec$ complies with the rank, i.e.,

$$\mathsf{rk}\,\mathfrak{h}' < \mathsf{rk}\,\mathfrak{h} \;\Longrightarrow\; \mathfrak{h}' \prec \mathfrak{h}\,, \qquad \text{for all } \mathfrak{h}, \mathfrak{h}' \in \mathsf{HF}$$

(ii) $\prec$ extends the strict partial ordering $\subsetneq$ among sets of the same rank, i.e.,

$$[(\mathsf{rk}\,\mathfrak{h}' = \mathsf{rk}\,\mathfrak{h}) \wedge (\mathfrak{h}' \subsetneq \mathfrak{h})] \;\Longrightarrow\; \mathfrak{h}' \prec \mathfrak{h}\,, \qquad \text{for all } \mathfrak{h}, \mathfrak{h}' \in \mathsf{HF}$$

satisfies (P1) and (P2) above. This is the case, for instance, for the ordering on HF induced by the ACKERMANN ENCODING, defined, for $\mathfrak{h} \in \mathsf{HF}$, by the following recursive equation

$$\mathbb{N}(\mathfrak{h}) = \sum_{\mathfrak{h}'\in\mathfrak{h}} 2^{\mathbb{N}(\mathfrak{h}')}\,. \tag{2.30}$$

Indeed, it is an easy matter to check that, for each $\mathfrak{h} \in \mathsf{HF}$, we have

- $\mathbb{N}(\mathfrak{h}') < \mathbb{N}(\mathfrak{h})$, for every $\mathfrak{h}' \in \mathfrak{h}$, and

- $\mathbb{N}(\mathfrak{h}'') < \mathbb{N}(\mathfrak{h})$, for every $\mathfrak{h}'' \subsetneq \mathfrak{h}$.

---

[14]We recall that HF denotes the family of the hereditarily finite sets.

[15]Exercise 2.26 asks to prove that, for every hereditarily finite set $\mathfrak{h}$, a variable $x_{\{\mathfrak{h}'\}}$ occurs in $\varphi_{\{\mathfrak{h}\}}$ if and only if h' belongs to the transitive closure of $\{\mathfrak{h}\}$.

By induction on the Ackermann code $\mathbb{N}(\mathfrak{h})$ of $\mathfrak{h} \in \mathsf{HF}$, it can be easily shown that each MLSP-formula $\varphi_{\{\mathfrak{h}\}}$ is satisfiable and it also fulfills the above faithfulness condition (FC). Hence, any hereditarily finite $\mathfrak{h}$ can be expressed by an MLSP-conjunction, via the variable $x_{\mathfrak{h}}$, as follows:

$$\begin{cases} x_{\emptyset} = x_{\emptyset} \setminus x_{\emptyset} & \text{if } \mathfrak{h} = \emptyset \\ \left(x_{\mathfrak{h}} = \bigcup_{\mathfrak{h}' \in \mathfrak{h}} x_{\{\mathfrak{h}'\}} \right) \wedge \bigwedge_{\mathfrak{h}' \in \mathfrak{h}} \varphi_{\{\mathfrak{h}'\}} & \text{otherwise.} \end{cases}$$

Alternatively, $\mathfrak{h} \in \mathsf{HF}$ could faithfully be expressed, via the variable $x_{\{\mathfrak{h}\}}$, by the following simpler formula

$$x_{\mathfrak{h}} \in x_{\{\mathfrak{h}\}} \wedge \varphi_{\{\mathfrak{h}\}},$$

which involves, however, an occurrence of the membership operator.

Finally, we observe that, by eliminating compound terms from $\varphi_{\{\mathfrak{h}\}}$, through the introduction of fresh auxiliary variables, until every literal contains just one compound term,[16] each formula $\varphi_{\{\mathfrak{h}\}}$ can be easily transformed into an equisatisfiable MLSP-conjunction $\varphi'_{\{\mathfrak{h}\}}$ of positive literals of the types (2.28). Thus, we have that every hereditarily finite set is expressible by a normalized MLSP-conjunction involving no membership.

Recapitulating, we have shown the following results:

Lemma 2.44. *Every hereditarily finite set is expressible in both MLSC and MLSuC, and by a normalized MLSP-conjunction involving no membership.*

2.4.3 Expressing Terms of the Form $\mathcal{P}^*(\{x_1, \ldots, x_k\})$

We show now that also terms of the form $\mathcal{P}^*(\{x_1, \ldots, x_k\})$ can be expressed by conjunctions of positive MLSP-literals of type (2.28) (thus, not involving the membership operator).

To begin with, consider a set $\mathcal{S} = \{s_1, s_2\}$. Then we have

$$\mathcal{P}^*(\mathcal{S}) = \mathcal{P}(s_1 \cup s_2) \setminus \left(\mathcal{P}(s_1 \setminus s_2) \cup \mathcal{P}(s_2 \setminus s_1) \right). \tag{2.31}$$

Indeed, if $s \in \mathcal{P}^*(\mathcal{S})$, plainly $s \subseteq s_1 \cup s_2$, $s \nsubseteq s_1 \setminus s_2$, and $s \nsubseteq s_2 \setminus s_1$, so that $s \in \mathcal{P}(s_1 \cup s_2) \setminus \left(\mathcal{P}(s_1 \setminus s_2) \cup \mathcal{P}(s_2 \setminus s_1) \right)$. Conversely, if $s \in \mathcal{P}(s_1 \cup s_2) \setminus \left(\mathcal{P}(s_1 \setminus s_2) \cup \mathcal{P}(s_2 \setminus s_1) \right)$, then $s \subseteq s_1 \cup s_2$ and $s \cap s_1 \neq \emptyset \neq s \cap s_2$, which yields $s \in \mathcal{P}^*(\mathcal{S})$, by the very definition of the operator $\mathcal{P}^*$. Hence, (2.31) follows.

Equation (2.31) readily generalizes to

$$\mathcal{P}^*(\mathcal{S}) = \mathcal{P}\left(\bigcup \mathcal{S} \right) \setminus \left(\bigcup_{s \in \mathcal{S}} \mathcal{P}\left(\bigcup \mathcal{S} \setminus s \right) \right), \tag{2.32}$$

for every set $\mathcal{S}$.

From (2.32), it follows easily that any term of the form $\mathcal{P}^*(\{x_1, \ldots, x_k\})$, with $k \geqslant 0$, can be expressed by the following MLSP-formula[17] via the variable x

$$x = \mathcal{P}\left(\bigcup_{i=1}^{k} x_i \right) \setminus \left(\bigcup_{j=1}^{k} \mathcal{P}\left(\left(\bigcup_{i=1}^{k} x_i \right) \setminus x_j \right) \right), \tag{2.33}$$

---

[16]See Section 2.1.2.

[17]For $k = 0$, the literal (2.33) simplifies to $x = \mathcal{P}(\emptyset)$, whereas for $k = 1$ it simplifies to $x = \mathcal{P}(x_1) \setminus \mathcal{P}(\emptyset)$.

in the sense that

$$\models x = \mathcal{P}\left(\bigcup_{i=1}^{k} x_i\right) \setminus \left(\bigcup_{j=1}^{k} \mathcal{P}\left(\left(\bigcup_{i=1}^{k} x_i\right) \setminus x_j\right)\right) \;\rightarrow\; x = \mathcal{P}^*(\{x_1, \ldots, x_k\}).$$

Finally, as before, by eliminating compound terms from (2.33), through the introduction of fresh auxiliary variables, until every literal contains just one compound term, the formula (2.33) can be easily transformed into an equisatisfiable MLSP-conjunction $\varphi'_{\{b\}}$ of positive literals of the types (2.28). In conclusion, we have proved the following result:

Lemma 2.45. *Every term of the form $\mathcal{P}^*(\{x_1, \ldots, x_k\})$ is expressible by normalized* MLSP-*conjunctions involving no membership.*

2.4.4 Expressing Two Relations about Cardinalities

We show that

(A) positive occurrences[18] of the cardinality comparison relation $|X| = |Y|$, and

(B) positive occurrences of the finiteness unary relation $Finite(X)$

are expressible in the extension MLSuC$^+$ of MLSuC with literals of the form $y = \biguplus x$.

Towards an expressiveness proof for (A), we first prove the following two preliminary lemmas.

Lemma 2.46. *Let A, B, C be sets such that:*

(i) $A \cap B = \emptyset$;

(ii) $C \subseteq A \otimes B$;

(iii) $\biguplus C = A \cup B$.

Then we have $|A| = |B|$.

Proof. To prove that $|A| = |B|$ holds, it is enough to exhibit a bijection $f \in B^A$. By (iii), for every $a \in A$ there exists a unique set $c \in C$ such that $a \in c$. Hence, from (i) and (ii), there must exist a unique set $b \in B$ such that $c = \{a, b\}$. We put $f(a) := b$. It is an easy matter to verify that the map f so defined is indeed a bijection from A onto B. Hence, $|A| = |B|$ holds. $\qquad\square$

Lemma 2.47. *For all sets X, Y, the following two conditions are equivalent:*

(a) *there exists a set Z such that*

    (a1) $Z \subseteq X \otimes (\{X\} \otimes Y)$, *and*

    (a2) $\biguplus Z = X \cup (\{X\} \otimes Y)$;

---

[18] An occurrence of a literal ℓ within a given (unquantified) formula φ is *positive* if the path to the root of the syntax tree of φ from the root of ℓ, deprived by its first node, contains an even number of nodes labeled by the negation symbol $\neg$. Otherwise, the occurrence is said to be negative. Roughly speaking, an occurrence of a literal ℓ in φ is positive if it is not preceded by a negation symbol when φ is transformed into conjunctive or disjunctive normal form.

(b) $|X| = |Y|$.

Proof. Let X, Y be any two sets. Then we have $X \cap (\{X\} \otimes Y) = \emptyset$. Indeed, if $s \in \{X\} \otimes Y$, then $X \in s$, and, therefore, it cannot be the case that $s \in X$, since otherwise the well-foundedness of membership would be disrupted.

Let us assume that there exists a set Z such that (a1) and (a2) hold. Upon putting

$$A := X, \qquad B := \{X\} \otimes Y, \qquad C := Z,$$

the hypotheses of Lemma 2.46 hold for A, B, C. Hence,

$$|X| = |A| = |B| = |\{X\} \otimes Y| = |Y|,$$

proving (b).

Conversely, assume now that (b) holds, namely, $|X| = |Y|$. Let f be a bijection in $(\{X\} \otimes Y)^X$, and put:

$$Z := \{\{s, f(s)\} \mid s \in X\}.$$

It is easy to check that conditions (a1) and (a2) are satisfied for the set Z just defined. $\square$

From Lemma 2.47, it follows readily that positive instances of the relation $|X| = |Y|$ can be expressed via the set variables x and y by the following formula:

$$z \subseteq x \otimes (\{x\} \otimes y) \ \wedge \ \biguplus z = x \cup (\{x\} \otimes y). \tag{2.34}$$

In view of the fact that the singleton is expressible by MLSuC-formulae, by (2.34) it follows that positive occurrences of the relation $|X| = |Y|$ can be expressed by MLSuC$^+$-formulae. Hence, we have proved:

Lemma 2.48. *Positive occurrences of the relation $|X| = |Y|$ can be expressed by* MLSuC$^+$- *formulae.*

Next we provide an expressiveness proof for (B).

Define recursively the sequence of sets $\{a_n\}_{n \in \omega}$, where

$$
\begin{aligned}
a_0 &:= \{\emptyset\} \\
a_{n+1} &:= \{\emptyset, a_n\},
\end{aligned}
\tag{2.35}
$$

and put

$$
\begin{aligned}
A_n &:= \{a_i \mid 0 \leqslant i \leqslant n\}, \quad \text{for every} \quad n \in \omega, \quad \text{and} \\
A_\infty &:= \{a_n \mid n \in \omega\}.
\end{aligned}
\tag{2.36}
$$

The following property holds:

Lemma 2.49. *For every set Y, the following conditions are equivalent:*

(a) $Y \subsetneq \{\emptyset\} \otimes (\{\emptyset\} \cup Y)$,

(b) $Y = A_n$, *for some* $n \geqslant 0$.

Proof. Here we prove only the implication (a) $\implies$ (b), and leave to the reader, as Exercise 2.33, the proof of the converse implication.

Let Y be such that $Y \subsetneq \{\emptyset\} \otimes (\{\emptyset\} \cup Y)$, and let $Z := \{\emptyset\} \otimes (\{\emptyset\} \cup Y)$. To begin with, we show that $Z \subseteq A_\infty$. We proceed by contradiction. If $Z \not\subseteq A_\infty$, we select a $\overline{z} \in Z \setminus A_\infty$ of minimal rank. As $\overline{z} \in \{\emptyset\} \otimes (\{\emptyset\} \cup Y)$, then $\overline{z} = \{\emptyset, \overline{y}\}$, for some $\overline{y} \in \{\emptyset\} \cup Y$. In fact, it must be the case that $\overline{y} \in Y$, since otherwise we would have $\overline{z} = \{\emptyset\} \in A_\infty$. Since rk $\overline{y} <$ rk $\overline{z}$ and $\overline{y} \in Z$, we have $\overline{y} \in A_\infty$, so that $\overline{y} = a_{\overline{k}}$, for some $\overline{k} \geqslant 0$. But then, we would have $\overline{z} = a_{\overline{k}+1} \in A_\infty$, a contradiction. Thus, we have $Z \subseteq A_\infty$.

Next we show that the implication

$$a_k \in Z \qquad \implies \qquad a_\ell \in Y, \text{ for every } 0 \leqslant \ell < k \qquad (2.37)$$

holds for every $k \geqslant 0$. We proceed again by contradiction. Assume that (2.37) is false for some $k \geqslant 0$, and let $\overline{k} \in \omega$ be the minimal k for which (2.37) does not hold. Hence, $\overline{k} > 0$ and $a_{\overline{k}} = \{\emptyset, a_{\overline{k}-1}\}$, with $a_{\overline{k}-1} \in Y$. By the minimality of $\overline{k}$ and since $Y \subsetneq Z$, we have that $a_{k'} \in Y$, for all $0 \leqslant k' < \overline{k} - 1$, and since $a_{\overline{k}-1} \in Y$, it follows that (2.37) holds for $\overline{k}$, a contradiction.

From (2.37), we have that $|Z \setminus Y| = 1$. Indeed, if $a_{k_1}, a_{k_2} \in Z \setminus Y$, with $k_1 \neq k_2$, then (2.37) would entail $\min(a_{k_1}, a_{k_2}) \in Y$, a contradiction. Thus, let $Z \setminus Y = \{a_{\overline{k}}\}$. From what we have shown above, it follows that

$$Z = \{a_i \mid 0 \leqslant i \leqslant \overline{k}\} = A_{\overline{k}} \text{ and } Y = \{a_i \mid 0 \leqslant i \leqslant \overline{k} - 1\} = A_{\overline{k}-1}.$$

This completes the proof that (a) implies (b). $\qquad\qquad\qquad\qquad\qquad\qquad\qquad\qquad\square$

By Lemma 2.49, the formula

$$|x| = |y| \wedge y \subsetneq (\{\emptyset\} \otimes (\{\emptyset\} \cup y))$$

expresses positive instances of the relation *Finite*(X) via the set variable x. Since the singleton can be expressed by MLSuC-formulae (Lemma 2.42) and positive occurrences of the cardinality relation $|X| = |Y|$ are expressible by MLSuC$^+$-formulae (Lemma 2.48), it follows that positive instances of the relation *Finite*(X) can be expressed by MLSuC$^+$-formulae. Hence, we have:

Lemma 2.50. *Positive occurrences of the relation Finite*(X) *can be expressed by* MLSuC$^+$-*formulae.*

In the following section, we discuss two applications of the expressiveness results proved here.

2.5 Two Applications

As application of some of the expressivity results presented in the previous section, we prove the undecidability of the satisfiability problem for MLSuC$^+$. In addition we show that MLSuC is not rank dichotomic, where *rank dichotomy* is a certain model-theoretic property related to the small model property.

2.5.1 Undecidability of MLSuC$^+$

Hilbert's Tenth Problem (HTP) is to find an algorithm that decides whether or not any given Diophantine equation (namely, a polynomial equation with integer coefficients) has an integer solution. It has been posed as the tenth problem in a list of 23 in a well-celebrated talk that David Hilbert addressed before the International Congress of Mathematicians in 1900 (cf. [Hil00]). HTP has been solved negatively by Yuri Matiyasevich in 1970 (cf. [Mat93, Mat93]), who built on the 21 years' long work of Martin Davis, Hilary Putnam, and Julia Robinson.[19]

The satisfiability problems for the extensions of MLSC and MLSuC with cardinality comparison literals of the form

$$|x| \leqslant |y|, \qquad |x| < |y|$$

have been proved undecidable in [CCP89], by reducing them to HTP. In fact, by using much the same reduction technique presented in [CCP89], it can be shown that undecidability is trigged as soon as one extends normalized MLSC- and MLSuC-conjunctions with *positive* literals of the following two types:

$$|x| = |y|, \qquad Finite(x). \tag{2.38}$$

Since by Lemmas 2.48 and 2.50 positive occurrences of both literals in (2.38) can be expressed by MLSuC$^+$-formulae, in the light of the above discussion, we have the following undecidability result:

Lemma 2.51. *The satisfibility problem for* MLSuC$^+$ *is undecidable.* ∎

Remark 2.52. Observe that positive occurrences of the literal $|x| = |y|$ can be also expressed by MLSuC-formulae extended with the standard union operator $\bigcup$ and the predicate isPartition(x), expressing that x is a partition. Indeed, $|x| = |y|$ is equisatisfiable with

$$z \subseteq x \otimes (\{x\} \otimes y) \wedge \text{isPartition}(z) \wedge \bigcup z = x \cup (\{x\} \otimes y),$$

where z is a fresh set variable (see Exercise 2.29).

Similarly, the literal $|x| = |y|$ can be expressed by MLSuC-formulae extended with the predicate isPartitionOf(x, y) expressing that x is a partition of y, namely, x is a set partition with set domain y. Indeed, $|x| = |y|$ is equisatisfiable with

$$z \subseteq x \otimes (\{x\} \otimes y) \wedge \text{isPartitionOf}(z, x \cup (\{x\} \otimes y)),$$

where z is a fresh set variable.

Thus, it follows that the satisfiability problem for the theories

- MLSuC extended with the standard union operator $\bigcup$ and the predicate isPartition(x),

- MLSuC extended with the predicate isPartitionOf(x, y)

is undecidable in both cases. ∎

---

[19]For a self-contained presentation of a solution to HTP, the reader is referred to [Dav73]; see also [OP16] for some very recent accounts on HTP and its extensions.

2.5.2 Rank Dichotomy

We introduce two model-theoretic properties, under the common name of RANK DICHOTOMY, which are variations of the small model property, as will be clear from Lemma 2.56 below.

Definition 2.53 (Rank dichotomy). *A fragment $\mathfrak{F}$ of the theory $\mathcal{S}$ enjoys the* RANK DICHOTOMY PROPERTY *(or, equivalently, it is* RANK DICHOTOMIC*) if there exists a computable function $g_{\mathfrak{F}}\colon \omega \to \omega$ (called* RANK DICHOTOMY FUNCTION *for $\mathfrak{F}$) such that, for every $\varphi \in \mathfrak{F}$, either*

- *φ admits a model M of infinite rank, namely, such that* rk $(M[\mathrm{Vars}(\varphi)]) \geqslant \omega$, *or*

- *every model M for φ, if any, has its rank bounded by $g_{\mathfrak{F}}(|\mathrm{Vars}(\varphi)|)$, namely,*

$$\mathrm{rk}\left(\bigcup M[\mathrm{Vars}(\varphi)]\right) \leqslant g_{\mathfrak{F}}(|\mathrm{Vars}(\varphi)|)\,. \qquad\blacksquare$$

Definition 2.54 (Strong rank dichotomy). *A fragment $\mathfrak{F}$ of the theory $\mathcal{S}$ enjoys the* STRONG RANK DICHOTOMY PROPERTY *(or it is* STRONGLY RANK DICHOTOMIC*) if there exists a computable function $h_{\mathfrak{F}}\colon \omega \to \omega$ (called* STRONG RANK DICHOTOMY FUNCTION *for $\mathfrak{F}$) such that, for every $\varphi \in \mathfrak{F}$, either*

- *φ admits an infinite model M, namely, such that $|\bigcup M[\mathrm{Vars}(\varphi)]| \geqslant \aleph_0$, or*

- *every model M for φ, if any, has its rank bounded by $h_{\mathfrak{F}}(|\mathrm{Vars}(\varphi)|)$.* $\qquad\blacksquare$

As the names suggest, if a theory is strongly rank dichotomic, then it is rank dichotomic. It is enough to observe that, for any model M of a given formula φ, we have

$$|\bigcup M[\mathrm{Vars}(\varphi)]| \geqslant \aleph_0 \quad \Longrightarrow \quad \mathrm{rk}\,(M[\mathrm{Vars}(\varphi)]) \geqslant \omega\,.$$

However, the converse does not hold, as shown in the following example.

Example 2.55. Consider the following toy fragment of the theory $\mathcal{S}$

$$\mathfrak{F}_0 := \left\{\, \bigcup x \subseteq \bigcup\bigcup x \,\wedge\, \bigcup x \neq \varnothing \,\wedge\, \mathit{Finite}(x) \,\right\},$$

containing the single formula

$$\varphi_0 \overset{\mathrm{Def}}{:=} \bigcup x \subseteq \bigcup\bigcup x \,\wedge\, \bigcup x \neq \varnothing \,\wedge\, \mathit{Finite}(x)\,.$$

Plainly, φ_0 is satisfiable only by models M such that $\bigcup Mx$ is infinite (and, therefore, such that Mx has infinite rank). For instance, the set assignment M_0 where

$$M_0 x := \left\{\, \{\varnothing, \varnothing^1, \varnothing^2, \dots\} \,\right\}$$

satisfies φ_0; in addition, M_0 is finite but has infinite rank. Hence $\mathfrak{F}_0$ is rank dichotomic, but not strongly rank dichotomic.

Notice, however, that the very similar theory

$$\mathfrak{F}_0' := \left\{\, y = \bigcup x \,\wedge\, y \subseteq \bigcup\bigcup x \,\wedge\, \bigcup x \neq \varnothing \,\wedge\, \mathit{Finite}(x) \,\right\}$$

is both rank dichotomic and strongly rank dichotomic. $\qquad\blacksquare$

The preceding example highlights that the property for a theory of being (strongly) rank dichotomic may depend on the form of its formulae. When investigating the rank dichotomy property for some fragment of set theory, it is therefore convenient, for uniformity, to restrict oneself only to theories whose formulae are normalized. We expect that the notions of rank dichotomy and of strong rank dichotomy coincide in the case of collections of normalized conjunctions of the theory $\mathcal{S}$.

Let $\mathfrak{F}$ be any fragment of our theory $\mathcal{S}$, and consider the subfragment $\widehat{\mathfrak{F}}$ of $\mathfrak{F}$, called the HEREDITARILY FINITE PART OF $\mathfrak{F}$, consisting of all the formulae in $\mathfrak{F}$ that do not admit any model of infinite rank, i.e.,

$$\widehat{\mathfrak{F}} := \left\{ \varphi \in \mathfrak{F} \mid \Phi \wedge \bigvee_{x \in \mathrm{Vars}(\varphi)} x \notin \mathsf{HF} \text{ is unsatisfiable} \right\},$$

where HF is an interpreted constant symbol such that $M(\mathsf{HF}) = \mathsf{HF}$, for every set assignment M. Likewise, consider the subfragment $\widehat{\widehat{\mathfrak{F}}}$ of $\mathfrak{F}$, called FINITE PART OF $\mathfrak{F}$, consisting of all the formulae in $\mathfrak{F}$ that do not admit any infinite model, i.e.,

$$\widehat{\widehat{\mathfrak{F}}} := \left\{ \varphi \in \mathfrak{F} \mid \Phi \wedge \bigvee_{x \in \mathrm{Vars}(\varphi)} \neg \mathit{Finite}(x) \text{ is unsatisfiable} \right\}.$$

Then the following results point up the relationship between rank dichotomy and the small model property.

Lemma 2.56. *For any fragment $\mathfrak{F}$ of the theory $\mathcal{S}$, we have:*

(a) *$\mathfrak{F}$ is rank dichotomic if and only if its hereditarily finite part $\widehat{\mathfrak{F}}$ enjoys the small model property;*

(b) *$\mathfrak{F}$ is strongly rank dichotomic if and only if its finite part $\widehat{\widehat{\mathfrak{F}}}$ enjoys the small model property.* ∎

As a consequence, we have:

Corollary 2.57. *For any fragment $\mathfrak{F}$ of the theory $\mathcal{S}$, we have:*

(a) *if $\mathfrak{F}$ is rank dichotomic, then its hereditarily finite part $\widehat{\mathfrak{F}}$ has a decidable hereditarily finite satisfiability probem;*

(b) *if $\mathfrak{F}$ is strongly rank dichotomic, then its finite part $\widehat{\widehat{\mathfrak{F}}}$ has a decidable hereditarily finite satisfiability probem.*

The proofs of Lemma 2.56 and its Corollary 2.57 are left to the reader as Exercise 2.31.

By Lemma 2.49, it follows quite easily that the theory MLSuC is not rank dichotomic.

Lemma 2.58. MLSuC *is not rank dichotomic.*

Proof. Consider the following formula

$$y \subsetneq \left(\{\varnothing\} \otimes (\{\varnothing\} \cup y) \right). \tag{2.39}$$

As already observed, singletons are expressible by MLSuC-formulae (see Lemma 2.42). Hence, the formula (2.39) can be rewritten as a normalized MLSuC-conjunction. Therefore Lemma 2.49 yields that (2.39) admits models of arbitrarily large finite ranks, while not admitting any model of infinite rank. Thus, the theory MLSuC is not rank dichotomic. □

A simple example of a strongly rank dichotomic theory is the fragment BST (see Exercise 2.32). In addition, as an application of the formative process technique, in Section 5.7 we shall prove that the theory MLSP is strongly rank dichotomic as well (see Lemma 5.17 and its Corollary 5.18).

EXERCISES

Exercise 2.1. *Complete the definition of $\mathcal{S}$-syntax tree given in Section 2.1.1, by extending it to $\mathcal{S}$-formulae too.*

Exercise 2.2. *The* height *of an $\mathcal{S}$-expression is the height of its syntax tree, namely, the length of the longest path in its syntax tree from the root to a leaf.*
    Provide a recursive definition of the height of $\mathcal{S}$-terms/$\mathcal{S}$-formulae.

Exercise 2.3. *Prove that the formulae*

$$\Phi_2 \overset{\text{Def}}{:=} x \times x \subseteq x \ \wedge \ x \neq \varnothing$$

$$\Phi_3 \overset{\text{Def}}{:=} x \otimes x \subseteq x \ \wedge \ x \neq \varnothing$$

admit only infinite models.

Exercise 2.4. *Let M be a set assignment over a collection V of set variables, Σ_M the Venn partition induced by M, and $\mathfrak{I}_M : V \to \mathcal{P}(\Sigma_M)$ the map defined by putting*

$$\mathfrak{I}_M(v) := \{\sigma \in \Sigma_M \mid \sigma \subseteq Mv\}, \qquad \text{for } v \in V.$$

Show that the set assignment induced by $\mathfrak{I}_M$ is just M.

Exercise 2.5. *Show that if an $\mathcal{S}$-formula is satisfied by a set assignment, then it is satisfied by some partition.*

Exercise 2.6. *Let Σ be a partition, and $M_{\mathfrak{I}}$ the set assignment induced by a given map $\mathfrak{I}: V \to \mathcal{P}(\Sigma)$ over a collection of set variables V. Show that, for $x, y, z \in V$,*

$$M_{\mathfrak{I}} \models x = y \setminus z \quad \text{if and only if} \quad \mathfrak{I}(x) = \mathfrak{I}(y) \setminus \mathfrak{I}(z).$$

Exercise 2.7. *Prove Lemma 2.21.*

Exercise 2.8. *Let $\beta \colon \Sigma \rightarrowtail \widehat{\Sigma}$ be a bijection from a partition Σ to another partition $\widehat{\Sigma}$, and let $\overline{\beta}$ and $\overline{\beta^{-1}}$ be, respectively, the image-maps related to β and β^{-1}. Prove that $\overline{\beta}$ is a bijection and that $\overline{\beta^{-1}} = \overline{\beta}^{-1}$ holds.*

Exercise 2.9. *Show that if a partition Σ is simulated by a second partition $\widehat{\Sigma}$, and $\widehat{\Sigma}$, in its turn, is simulated by a third partition $\widehat{\widehat{\Sigma}}$, then Σ is simulated by $\widehat{\widehat{\Sigma}}$.*

Exercise 2.10. *Show that if a partition Σ is imitated by a second partition $\widehat{\Sigma}$, which, in its turn, is imitated by a third partition $\widehat{\widehat{\Sigma}}$, then Σ is imitated by $\widehat{\widehat{\Sigma}}$.*

Exercise 2.11. *Prove that, when $\widehat{\Sigma}$ and Σ are $\mathcal{P}$-partitions, condition (iii) of Definition 2.26 is implied by condition (i).*

Exercise 2.12. *Prove Lemma 2.30.*

Exercise 2.13. *Prove Lemma 2.35.*

Exercise 2.14. *Let F be a 3-SAT instance of the form (2.22) and Φ_F^* its corresponding BST-conjunction (2.23). Show in detail that F is propositionally satisfiable if and only if the BST-conjunction Φ_F^* is satisfiable by a set model.*

Exercise 2.15. *Show that the satisfiability problem for the subtheory B_0 of BST consisting of the conjunctions of literals of the following three types*

$$x = y \cup z, \qquad x \cap y = \varnothing, \qquad x \neq y$$

is NP-complete.

Exercise 2.16. *Show that the satisfiability problem for the subtheory B_1 of BST consisting of the conjunctions of literals of the following two types*

$$x = y \setminus z, \qquad x \neq y$$

is NP-complete.

Exercise 2.17. *Let B_2 be the subtheory of BST consisting of the conjunctions of literals of the following three types*

$$x = y \cup z, \qquad x = y \cap z, \qquad x \neq y$$

which do not involve the constant $\varnothing$ (thus, x, y, z in the literals above can stand only for set variables).
   Show that

(a) *it is not possible to express the set difference operator '$\setminus$' in B_2, in the sense that, given any set variables x, y, z, there is no satisfiable B_1-conjunction Φ such that the implication*

$$\Phi \;\rightarrow\; x = y \setminus z$$

   is true (i.e., satisfied by every set assignment over the variables $\mathrm{Vars}(\Phi) \cup \{x, y, z\}$);

(b) *the satisfiability problem for B_2 is NP-complete.*

[Hint for point (a): First show that B_2 cannot express the empty set '$\varnothing$'.]

Exercise 2.18. *Prove that the singleton operator in not expressible in MLS.*

Exercise 2.19. Let $\psi_\times(x,y)$ be the conjunction of the literals (2.24). Prove that, for any set X, there exists a set assignment M_X, over the variables of $\psi_\times(x,y)$, that satisfies $\psi_\times(x,y)$, and such that $Mx = X$ and $My = \{Mx\}$ hold.

Exercise 2.20. Let $\psi_\otimes(x,y)$ be the conjunction of the literals (2.25). Prove that the ML-SuC-conjunction $\psi_\otimes(x,y)$ expresses the singleton operation.

Exercise 2.21. Prove that the following MLSuC-formula

$$\left(x \in y \subseteq y' \wedge y \in y' \wedge y' \in y' \otimes y'\right) \rightarrow y \in y' \times y'$$

is true.

Exercise 2.22. Prove rigorously that the hereditarily finite sets can be expressed in MLSS and in MLSuC.

Exercise 2.23. Show that every representing formula $\varphi_{\{\mathfrak{h}\}}$, for $\mathfrak{h} \in \mathsf{HF}$, is satisfiable and enjoys the faithfulness condition.

Exercise 2.24. Prove that any total ordering $\prec$ of HF, such that the conditions

  1. $\mathrm{rk}\,\mathfrak{h}' < \mathrm{rk}\,\mathfrak{h} \implies \mathfrak{h}' \prec \mathfrak{h}$, and

  2. $[(\mathrm{rk}\,\mathfrak{h}' = \mathrm{rk}\,\mathfrak{h}) \wedge (\mathfrak{h}' \subsetneq \mathfrak{h})] \implies \mathfrak{h}' \prec \mathfrak{h}$

hold for all $\mathfrak{h}, \mathfrak{h}' \in \mathsf{HF}$, satisfies properties (P1) and (P2) in Section 2.4.2.

Exercise 2.25. Prove that the ordering on HF induced by the Ackermann encoding (2.30) satisfies properties (P1) and (P2) in Section 2.4.2.

Exercise 2.26. Show that, for all $\mathfrak{h}, \mathfrak{h}' \in \mathsf{HF}$, we have

$$x_{\{\mathfrak{h}'\}} \quad occurs\ in\ \ \varphi_{\{\mathfrak{h}\}} \qquad \Longleftrightarrow \qquad \mathfrak{h}' \in \mathsf{TrCl}(\{\mathfrak{h}\}).$$

Exercise 2.27. Prove that, for every set $\mathcal{S}$, we have

$$\mathcal{P}^*(\mathcal{S}) = \mathcal{P}\left(\bigcup \mathcal{S}\right) \setminus \left(\bigcup_{s \in \mathcal{S}} \mathcal{P}\left(\bigcup \mathcal{S} \setminus s\right)\right).$$

Exercise 2.28. Show that the map f defined in the proof of Lemma 2.46 is indeed a bijection from A onto B.

Exercise 2.29. Let A, B, C be sets such that

  - $A \cap B = \emptyset$,
  - $C \subseteq A \otimes B$,
  - C is a partition, and
  - $\bigcup C = A \cup B$.

Prove that $|A| = |B|$.

Exercise 2.30. Provide a fragment of the theory $\mathcal{S}$ that is rank dichotomic, but not strongly rank dichotomic, and whose formulae do not involve the union set operator $\bigcup$.

Exercise 2.31. Prove Lemma 2.56 and Corollary 2.57.

Exercise 2.32. Show that the theory BST is strongly rank dichotomic.

Exercise 2.33. Complete the proof of Lemma 2.49 (namely, prove that (b) $\implies$ (a)).

Chapter 3

Formative Processes

As seen in the previous chapter, in the case of S-formulae, classical *satisfiability by set assignments* is equivalent to *satisfiability by partitions* (cf. Lemma 2.18). In fact, we can further restrict ourselves to *satisfiability by $\mathcal{P}$-partitions*.

Although less natural, satisfiability by partitions turns out to be very useful in relation to the decision problem in set theory. This is particularly true for fragments involving operators and relators able to constrain the structure of the elements belonging to specific blocks of a satisfying partition, such as the set operators $\mathcal{P}$, $\bigcup$, $\times$, $\otimes$, and, to a lesser extent, the membership relator $\in$ and finite enumerations $\{\cdot, \ldots, \cdot\}$.

The approach of satisfying partitions has already been illustrated in the previous chapter, in connection with the decidability of the theory BST (involving only the operators $\cup$, $\setminus$, and the relator $=$). However, in that case, only a small part of the full apparatus of that approach was necessary, as the only constraints on a satisfying partition concerned the number of its blocks.

As soon as additional constructs, besides those of BST, come into play, the satisfiability-by-partitions approach requires more specialized techniques. In some cases, as for the MLS or MLSS theories, one can still rely on the standard satisfiability-by-assignments approach, which, due to its simplicity, is considerably preferable in such cases (cf. [FOS80] and [CFO89, Chapters 5 and 6]). However, in presence of more expressive operators such as $\mathcal{P}$ and $\bigcup$, resorting to the *formative process technique* in connection with the satisfiability-by-partitions approach becomes unavoidable.

In view of its application to the satisfiability problems for MLSSP and MLSSPF, to be addressed in the second part of the book, in this chapter we introduce the basic elements of the formative process technique, especially tailored to MLSP-like fragments, namely, the fragments MLSP, MLSSP, and MLSSPF. Our general strategy will be to establish a small model property for the theory MLSSP, and then reduce the satisfiability problem for MLSSPF to that of MLSSP.

Thus, given a satisfiable MLSSP-formula Φ and a satisfying $\mathcal{P}$-partition Σ, it will be of primary importance to be able to 'simplify' Σ, so that the resulting $\mathcal{P}$-partition $\widehat{\Sigma}$ has a finite bounded rank, while still satisfying Φ.

We have already provided sufficient conditions in order that a particular $\mathcal{P}$-partition $\widehat{\Sigma}$ satisfies all the MLSSP-formulae that hold for another $\mathcal{P}$-partition Σ. Specifically, we have proved that when a $\mathcal{P}$-partition $\widehat{\Sigma}$ *simulates* another $\mathcal{P}$-partition Σ, then $\widehat{\Sigma}$ satisfies all the MLSSP-formulae satisfied by Σ.

© Springer International Publishing AG, part of Springer Nature 2018
D. Cantone and P. Ursino, *An Introduction to the Technique of Formative Processes in Set Theory*, https://doi.org/10.1007/978-3-319-74778-1_3

The notion of *simulation* has been slightly specialized to that of *imitation* in such a way that if a $\mathcal{P}$-partition $\widehat{\Sigma}$ *imitates* another $\mathcal{P}$-partition Σ, then it also simulates it. Thus, a small model property for **MLSSP** is proved once one shows that:

Every $\mathcal{P}$-partition is imitated by a 'small' $\mathcal{P}$-partition. $(\star)$

To prove $(\star)$, we need to carry on a careful analysis of the construction process of $\mathcal{P}$-partitions.

Similarly to how the powerset operator $\mathcal{P}$ allows one to build, starting from the empty layer, the distinct layers of the von Neumann standard cumulative hierarchy (cf. Chapter 1), the intersecting powerset operator $\mathcal{P}^*$ can be used as the basic constructor for assembling any transitive partition (in particular, $\mathcal{P}$-partitions) starting from the empty 'node' of an underlying *syllogistic board*.

Syllogistic boards are bipartite directed graphs whose vertices are of two types, namely, *places* and *nodes*, where nodes are sets of places. Places and nodes of a syllogistic board are intended to represent, respectively, the blocks of the $\mathcal{P}$-partition one is about to assemble, and the corresponding sets of blocks. Initially, all places are just empty repositories that will get new elements step after step, until they reach their final configurations.

Syllogistic boards can be seen as a kind of *flow network*, having places as *sinks* and nodes as *sources*. The *flow* is formed by all the members in the domain of the $\mathcal{P}$-partition under construction. At any step, a node B can distribute among its targets only those elements belonging to $\mathcal{P}^*(B)$ that have not already been distributed in previous steps. The sequence of construction steps of a given $\mathcal{P}$-partition constitutes a *formative process* for it.

Some steps in a formative process are particularly relevant, as they correspond to special events in the construction (such as, for instance, the first time some elements flow through a connection). By suitably synchronizing the special events of different processes, it is possible to define a *shadowing* relationship among formative processes, in such a way that when a formative process is shadow of another one, then the final partitions of the two processes imitate one another.

What use can we make of formative processes for the simplification of $\mathcal{P}$-partitions?

Roughly speaking, starting from a formative process $\mathcal{H}$ of a given $\mathcal{P}$-partition, one can suitably prune its sequence of steps from 'irrelevant' steps (depending on the theory of which one is investigating the satisfiability problem), ending up in a shorter subsequence $S_{\mathcal{H}}$ of the *salient steps* of $\mathcal{H}$. The subsequence $S_{\mathcal{H}}$ enjoys the following two properties: (i) its length is bounded by a computable function of the size of the partition Σ, and (ii) there exists a formative process $\widehat{\mathcal{H}}$ whose sequence of steps is $S_{\mathcal{H}}$ and is such that $\widehat{\mathcal{H}}$ is a shadow process of $\mathcal{H}$. In the light of the above arguments, the final $\mathcal{P}$-partition of $\widehat{\mathcal{H}}$ satisfies exactly the same normalized conjunctions as the initial $\mathcal{P}$-partition. As remarked, such applications will be the main subject of the second part of the book.

3.1 A Gentle Introduction

Our next step consists of laying the foundations for techniques of partition manipulation, that maintain invariant the collection of the MLSSP-formulae satisfied by the partitions.

To that end, we need to characterize the 'recipe' of how any given transitive partition is constructed, and also to identify the *essential* instructions relative to MLSP-like theories of our interest: these must be maintained in any 'recipe' variation, in the very same order, whereas other subsequences of *inessential* instructions could be discarded or inserted, according to one's need. However, special care must be taken in order to produce 'realizable recipes', namely, 'recipes' which may lead to the actual construction of a partition.

It turns out that the proper construction granularity is achieved using the intersecting powerset operator $\mathcal{P}^*$ as basic constructor for partitions assemblage. Thus, any transitive partition can be built, starting from the empty partition, by a (possibly transfinite) sequence of $\mathcal{P}^*$-*enlargements*, in the sense that we are going to illustrate.

We agree that a partition Σ' is a (STRICT) ENLARGEMENT of another partition Σ (or, using a more consolidated terminology, Σ' FRAMES Σ; cf. Section 3.3) when $\Sigma' \neq \Sigma$ and there exists an *injective* map $\sigma \mapsto \sigma'$, from Σ into Σ', such that $\sigma \subseteq \sigma'$, for each $\sigma \in \Sigma$. In other words, Σ' is an enlargement of Σ when $\Sigma' \neq \Sigma$ and Σ' is the result of adding new blocks to Σ and/or replacing any blocks in Σ with proper supersets (still maintaining mutual disjointness).

Example 3.1. Given the partition

$$\Sigma_0 := \{\{0,1\},\{2,3\}\}$$

(where natural numbers are encoded *à la* von Neumann), then

$$\{\{0,1\},\{2,3\},\{4,5\}\}$$

is an enlargement of Σ_0. Further,

$$\{\{0,1,4\},\{2,3,5\}\} \qquad \text{and} \qquad \{\{0,1,4\},\{2,3\},\{5,6,7\}\}$$

are enlargements of Σ_0 as well, whereas

$$\{\{0,1,2,3,4\},\{5,6,7\}\}$$

is not. ∎

We say that a partition Σ' is a $\mathcal{P}^*$-ENLARGEMENT of Σ when

1. Σ' is an enlargement of Σ, and

2. $\bigcup \Sigma' \setminus \bigcup \Sigma \subseteq \mathcal{P}^*(\Gamma)$, for some $\Gamma \subseteq \Sigma$.

Point 2 says that all the new elements in the enlargement, namely, the members of $\bigcup \Sigma' \setminus \bigcup \Sigma$, not only must be formed with elements already present in the domain of Σ, but also they have to belong to the same set $\mathcal{P}^*(\Gamma)$, for some $\Gamma \subseteq \Sigma$.

Example 3.1 (cntd.). The two partitions

$$\{\{0,1\},\{2,3,\{1\}\}\} \qquad \text{and} \qquad \{\{0,1\},\{2,3,\{0,2\}\},\{\{1,2\},\{0,3\}\}\}$$

are $\mathcal{P}^*$-enlargements of Σ_0, whereas

$$\{\{0,1\},\{2,3,\{0,1\}\}\} \quad \text{and} \quad \{\{0,1\},\{2,3,\{0\}\}\}$$

are not (why?).   ∎

As will be shown in the Trace Theorem, it turns out that, for every transitive partition Σ (and, *a fortiori*, for every $\mathcal{P}$-partition), there exists a possibly transfinite sequence $\left(\Pi_\mu\right)_{\mu \leqslant \xi}$ of (transitive) partitions such that:

initial step: Π_0 is the empty partition, i.e., $\Pi_0 = \emptyset$;

successor step: each partition $\Pi_{\nu+1}$ is a $\mathcal{P}^*$-enlargement of Π_ν, for every $\nu < \xi$;

limit step: for every infinite limit ordinal $\lambda \leqslant \xi$, Π_λ is the *minimal* enlargement of all partitions Π_ν, with $\nu < \lambda$;

final step: Π_ξ is the partition Σ.

Any sequence $\left(\Pi_\mu\right)_{\mu \leqslant \xi}$ of partitions, whose construction takes place according to the above four steps, will be called a FORMATIVE PROCESS for the transitive partition Σ.[1] Notice that in this context the operator $\mathcal{P}^*$ plays much the same role as that of the power set operator $\mathcal{P}$ plays in the construction of the von Neumann standard cumulative hierarchy of sets.

3.1.1   Constructing a Formative Process

It is instructive to track the step-by-step construction of a formative process for the following transitive partition:

$$\Sigma_0 = \left\{ q_0^{(\bullet)}, q_1^{(\bullet)}, \ldots, q_4^{(\bullet)} \right\} \tag{3.1}$$

(depicted in Figure 3.1), where

$$q_0^{(\bullet)} = \{\emptyset\}, \qquad\qquad q_1^{(\bullet)} = \{\emptyset^1, \{\emptyset, \emptyset^2, \{\emptyset, \emptyset^1\}\}\}, \quad q_2^{(\bullet)} = \{\emptyset^2, \{\emptyset, \emptyset^1\}\},$$
$$q_3^{(\bullet)} = \{\emptyset^3, \{\emptyset^2, \{\emptyset, \emptyset^1\}\}\}, \quad q_4^{(\bullet)} = \{\{\emptyset, \emptyset^2\}\}$$

(we are using again the nesting notation $\emptyset^n$ of Example 1.12).

We need to interpolate all the intermediate partitions between the initial partition Π_0 and the final one (namely, the partition Σ_0). It is reasonable to assume that the elements in the domain $\bigcup \Sigma_0$ of Σ_0 will enter the intermediate partitions in any order compatible with the rank. Thus, we order the eight elements in $\bigcup \Sigma_0$ by their rank

$$\bigcup \Sigma_0 = \{\ \underbrace{\emptyset}_{0}, \underbrace{\emptyset^1}_{1}, \underbrace{\emptyset^2, \{\emptyset, \emptyset^1\}}_{2}, \underbrace{\emptyset^3, \{\emptyset, \emptyset^2\}, \{\emptyset^2, \{\emptyset, \emptyset^1\}\}, \{\emptyset, \emptyset^2, \{\emptyset, \emptyset^1\}\}}_{3}\ \}$$

---

[1] In connection with a formative process $\left(\Pi_\mu\right)_{\mu \leqslant \xi}$ and in order to adhere to a later convention, in general we shall use ν and μ to denote ordinals in the ranges $0 \leqslant \nu < \xi$ and $0 \leqslant \mu \leqslant \xi$, respectively.

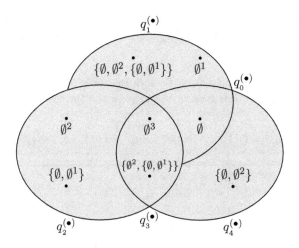

Figure 3.1: The transitive partition $\Sigma_0 = \{q_0^{(\bullet)}, q_1^{(\bullet)}, \ldots, q_4^{(\bullet)}\}$.

and process them accordingly.

The blocks of the μ-th partition Π_μ in the formative process $(\Pi_\mu)_{\mu \leqslant \xi}$ we are about to construct will be the non-null sets among

$$q_0^{(\mu)}, \; q_1^{(\mu)}, \; \ldots, \; q_4^{(\mu)},$$

where, initially, we have

$$q_0^{(0)} = q_1^{(0)} = \ldots = q_4^{(0)} = \emptyset,$$

and, accordingly,

$$\Pi_0 = \{q_i^{(0)} \mid q_i^{(0)} \neq \emptyset, \text{ for } i = 0, 1, \ldots, 4\} = \emptyset.$$

At step $\nu = 0, 1, \ldots, 5$, while processing the element $e_\nu \in \bigcup \Sigma_0$, we find the unique $\Gamma_\nu \subseteq \Pi_\nu$ such that $e_\nu \in \mathcal{P}^*(\Gamma_\nu)$. Besides the element e_ν, all remaining elements in $(\mathcal{P}^*(\Gamma_\nu) \setminus \bigcup \Pi_\nu) \cap \bigcup \Sigma_0$ could be also added to their respective blocks at this same step. If this happens consistently at any step of the construction, the final formative process will be said to be GREEDY (otherwise it is said to be WEAK).

Step $\nu = 0$, $e_1 = \emptyset$: The unique $\Gamma_0 \subseteq \Pi_0$ such that $\emptyset \in \mathcal{P}^*(\Gamma_0)$ is $\Gamma_0 := \emptyset$. We have $\mathcal{P}^*(\emptyset) = \{\emptyset\}$; thus, at this step we add the element $\emptyset$ to the set $q_0^{(0)}$, obtaining

$$q_0^{(1)} = \{\emptyset\}, \quad q_1^{(1)} = \ldots = q_4^{(1)} = \emptyset \qquad \text{and} \qquad \Pi_1 = \{q_0^{(1)}\}.$$

Step $\nu = 1$, $e_2 = \emptyset^1$: The unique $\Gamma_1 \subseteq \Pi_1$ such that $\emptyset^1 \in \mathcal{P}^*(\Gamma_1)$ is $\Gamma_1 := \{q_0^{(1)}\}$. We have $\mathcal{P}^*(\{q_0^{(1)}\}) = \{\emptyset^1\}$; thus, at this step we add the element $\emptyset^1$ to the set $q_1^{(1)}$, obtaining

$$q_0^{(2)} = \{\, \emptyset \,\}, \quad q_1^{(2)} = \{\, \emptyset^1 \,\}, \quad q_2^{(2)} = q_3^{(2)} = q_4^{(2)} = \emptyset \qquad \text{and} \qquad \Pi_2 = \{\, q_0^{(2)}, q_1^{(2)} \,\}.$$

Step $\nu = 2$, $e_3 = \emptyset^2$: The unique $\Gamma_2 \subseteq \Pi_2$ such that $\emptyset^2 \in \mathcal{P}^*(\Gamma_2)$ is $\Gamma_2 := \{\, q_1^{(2)} \,\}$. We have $\mathcal{P}^*(\{\, q_1^{(2)} \,\}) = \{\, \emptyset^2 \,\}$; thus, at this step we add the element $\emptyset^3$ to the set $q_2^{(2)}$, obtaining

$$q_0^{(3)} = \{\, \emptyset \,\}, \quad q_1^{(3)} = \{\, \emptyset^1 \,\}, \quad q_2^{(3)} = \{\, \emptyset^2 \,\}, \quad q_3^{(3)} = q_4^{(3)} = \emptyset$$

and

$$\Pi_3 = \{\, q_0^{(3)}, q_1^{(3)}, q_2^{(3)} \,\}.$$

Step $\nu = 3$, $e_4 = \{\, \emptyset, \emptyset^1 \,\}$: The unique $\Gamma_3 \subseteq \Pi_3$ such that $\{\, \emptyset, \emptyset^1 \,\} \in \mathcal{P}^*(\Gamma_3)$ is $\Gamma_3 := \{\, q_0^{(3)}, q_1^{(3)} \,\}$. We have $\mathcal{P}^*(\{\, q_0^{(3)}, q_1^{(3)} \,\}) = \{\, \{\, \emptyset, \emptyset^1 \,\} \,\}$; thus, at this step we add the element $\{\, \emptyset, \emptyset^1 \,\}$ to the set $q_2^{(3)}$, obtaining

$$q_0^{(4)} = \{\, \emptyset \,\}, \quad q_1^{(4)} = \{\, \emptyset^1 \,\}, \quad q_2^{(4)} = \{\, \emptyset^2, \{\, \emptyset, \emptyset^1 \,\} \,\}, \quad q_3^{(4)} = q_4^{(4)} = \emptyset$$

and

$$\Pi_4 = \{\, q_0^{(4)}, q_1^{(4)}, q_2^{(4)} \,\}.$$

Step $\nu = 4$, $e_5 = \emptyset^3$: The unique $\Gamma_4 \subseteq \Pi_4$ such that $\emptyset^3 \in \mathcal{P}^*(\Gamma_4)$ is $\Gamma_4 := \{\, q_2^{(4)} \,\}$. We have

$$\mathcal{P}^*(\{\, q_2^{(4)} \,\}) = \{\, \emptyset^3, \{\, \{\, \emptyset, \emptyset^1 \,\} \,\}, \{\, \emptyset^2, \{\, \emptyset, \emptyset^1 \,\} \,\} \,\};$$

at this step we can assign two elements, namely, $\emptyset^3$ and $\{\, \emptyset^2, \{\, \emptyset, \emptyset^1 \,\} \,\}$, and since we intend to construct a *greedy* formative process, we add both of them to the set $q_3^{(4)}$, obtaining

$$q_0^{(5)} = \{\, \emptyset \,\}, \ q_1^{(5)} = \{\, \emptyset^1 \,\}, \ q_2^{(5)} = \{\, \emptyset^2, \{\, \emptyset, \emptyset^1 \,\} \,\}, \ q_3^{(5)} = \{\, \emptyset^3, \{\, \emptyset^2, \{\, \emptyset, \emptyset^1 \,\} \,\} \,\},$$
$$q_4^{(5)} = \emptyset$$

and

$$\Pi_5 = \{\, q_0^{(5)}, q_1^{(5)}, q_2^{(5)}, q_3^{(5)} \,\}.$$

Step $\nu = 5$, $e_6 = \{\, \emptyset, \emptyset^2 \,\}$: The unique $\Gamma_5 \subseteq \Pi_5$ such that $\{\, \emptyset, \emptyset^2 \,\} \in \mathcal{P}^*(\Gamma_5)$ is $\Gamma_5 := \{\, q_0^{(5)}, q_2^{(5)} \,\}$. We have

$$\mathcal{P}^*(\{\, q_0^{(5)}, q_2^{(5)} \,\}) = \{\, \{\, \emptyset, \emptyset^2 \,\}, \{\, \emptyset, \{\, \emptyset, \emptyset^1 \,\} \,\}, \{\, \emptyset, \emptyset^2, \{\, \emptyset, \emptyset^1 \,\} \,\} \,\};$$

at this step we can assign two elements, namely, $\{\, \emptyset, \emptyset^2 \,\}$ and $\{\, \emptyset, \emptyset^2, \{\, \emptyset, \emptyset^1 \,\} \,\}$; we add the former to $q_4^{(5)}$ and the latter to $q_1^{(5)}$, thus obtaining

$$q_0^{(6)} = \{\emptyset\}, \; q_1^{(6)} = \{\emptyset^1, \{\emptyset, \emptyset^2, \{\emptyset, \emptyset^1\}\}\}, \; q_2^{(6)} = \{\emptyset^2, \{\emptyset, \emptyset^1\}\},$$
$$q_3^{(6)} = \{\emptyset^3, \{\emptyset^2, \{\emptyset, \emptyset^1\}\}\}, \; q_4^{(6)} = \{\{\emptyset, \emptyset^2\}\}$$

and

$$\Pi_6 = \{q_0^{(6)}, q_1^{(6)}, q_2^{(6)}, q_3^{(6)}, q_4^{(6)}\} = \{q_0^{(\bullet)}, q_1^{(\bullet)}, q_2^{(\bullet)}, q_3^{(\bullet)}, q_4^{(\bullet)}\} = \Sigma_0.$$

It is an easy matter to check that $\Pi_{\nu+1}$ is a $\mathcal{P}^*$-enlargement of Π_ν, for every $\nu = 0, 1, \ldots, 5$. Hence,

$$\left(\Pi_\mu\right)_{\mu \leqslant 6} \qquad\qquad (3.2)$$

is a greedy formative process for the transitive partition Σ_0.

3.1.2 Syllogistic Boards

From a closer analysis of the six construction steps of the above formative process $\left(\Pi_\mu\right)_{\mu \leqslant 6}$, it emerges that, at each step $0 \leqslant \nu \leqslant 5$, a non-null subset Γ_ν of the current partition Π_ν is selected, and some of the elements in $\mathcal{P}^*(\Gamma_\nu)$, new to Π_ν, are distributed among some prospect blocks $q_i^{(\nu)}$. It turns out that such subsets Γ_ν and prospect blocks $q_i^{(\nu)}$ are not chosen "randomly," but follow a certain pattern (depending on the initial partition Σ_0) which can be cast into a graph structure to be called SYLLOGISTIC BOARD (or SYLLOGISTIC Π-BOARD, when one wants to emphasize the set of places Π).

The most natural candidates for vertices of the syllogistic board induced by Σ_0 are, in our case, the objects $q_0, \ldots, q_4$ (which, appropriately superscripted, are used to denote the values of the prospect blocks at each stage) and their subsets. We therefore assume, without loss of any generality, that the q_i's are chosen in such a way that no q_i can equal a set of q_j's. The q_i's are called PLACES, while their subsets are called NODES. The collection of places of a syllogistic board is often denoted by Π, so that $\mathcal{P}(\Pi)$ will be the collection of its nodes. There is an edge $B \to q$ from a node B to a place q, in which case we say that q is a TARGET of B, just when the block $q^{(\bullet)}$ draws elements from $\mathcal{P}^*(B^{(\bullet)})$ (where we are putting $B^{(\bullet)} := \{p^{(\bullet)} \mid p \in B\}$), i.e., when $q^{(\bullet)} \cap \mathcal{P}^*(B^{(\bullet)}) \neq \emptyset$. In addition, we agree that, from every place q into every node B such that $q \in B$, there is an edge $q \to B$. As a result, a syllogistic board is a bipartite directed graph.

The syllogistic Π-board $\mathcal{G}$ induced by Σ_0, with $\Pi = \{q_0, \ldots, q_4\}$, is reported in Figure 3.3(a) where, for convenience, edges issuing from places have not been drawn, as they can be easily figured out from the diagram, and only nodes with targets have been drawn.

As remarked, the syllogistic Π-board induced by a transitive partition Σ (with set of places Π) can be seen as a kind of *flow network*, where sets constitute the flowing material. In this view, nodes are the *sources* of the network, whereas places are the *sinks*. At any step, each place is labeled by a set and, consequently, each node is labeled by the collection of the sets labeling its places; however, to ease our explanation, in what follows we identify each vertex with its label. A formative process for Σ is then generated by a sequence of *admissible flow-distribution steps* in the corresponding Π-board, where:

1. initially all places are empty;

2. a node B is *active* when it can distribute some 'new' elements among its targets, namely, when $(\mathcal{P}^*(B) \setminus \bigcup \Pi) \cap \bigcup \Sigma \neq \emptyset$ (the set $(\mathcal{P}^*(B) \setminus \bigcup \Pi) \cap \bigcup \Sigma$ of 'new' elements is the *admissible flow from B*);

3. an edge $B \to q$ issuing from an active node B is *active* when there are 'new' elements in $\mathcal{P}^*(B)$ that can flow into q without disrupting the set inclusion $q \subseteq q^{(\bullet)}$;

4. an *admissible flow-distribution step* from a node B requires that B is active, the admissible flow from it is distributed only through active edges, and, at the end of the distribution, the set inclusion $q \subseteq q^{(\bullet)}$ is fulfilled for each target of B.

Figure 3.2 exemplifies a generic admissible flow-distribution step. In addition, Figures 3.4 and 3.5 illustrate, respectively, the sequence of admissible flow distributions for the construction of our transitive partition Σ_0, and the final configuration of the induced Π-graph. Places are greyed out when they are empty. In addition, nodes and edges are greyed out when they are inactive. Finally, active edges are labeled with the sets flowing through them.

The sequence $(A_\nu)_{\nu < 6}$ of nodes

$$\emptyset, \ \{q_0\}, \ \{q_1\}, \ \{q_0, q_1\}, \ \{q_2\}, \ \{q_0, q_2\}$$

used in the flow-distribution steps of the formative process $(\Pi_\mu)_{\mu \leqslant 6}$ for Σ_0 depicted in Figure 3.4 forms the *trace* of $(\Pi_\mu)_{\mu \leqslant 6}$.

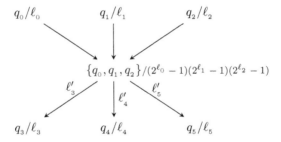

Figure 3.2: An admissible flow-distribution step. The blocks corresponding to the places q_0, q_1, q_2 contain, respectively, ℓ_0, ℓ_1, ℓ_2 elements. The node $\{q_0, q_1, q_2\}$ can therefore distribute up to $(2^{\ell_0} - 1) \cdot (2^{\ell_1} - 1) \cdot (2^{\ell_2} - 1)$ elements among its targets: ℓ'_3 elements are assigned to the target q_3, ℓ'_4 to the target q_4, and ℓ'_5 to the target q_5.

Remark 3.2. Syllogistic boards can also be represented in *contracted form*, to better grasp their underlying network structure. In contracted form, we distinguish two kinds of vertices:

1. *vertices labeled by places*: these have a twofold nature, as genuine places and as singleton nodes; and

2. *unlabeled places*, which stand for non-singleton nodes.

Denoting by "•" any unlabeled node, we therefore have:

- edges of type $q_i \rightarrow q_j$, where q_i and q_j are places, which stand for $\{q_i\} \rightarrow q_j$;

- edges of the form $q_i \rightarrow$ •, which stand for $q_i \rightarrow B$, where B is the set of all starting points of the edges entering the unlabeled vertex of $q_i \rightarrow$ •; and

- edges of type • $\rightarrow q_i$, which stand for $B \rightarrow q_i$, where B is the set of all starting points of the edges entering the unlabeled vertex of • $\rightarrow q_i$.

Notice that a syllogistic board in contracted form contains no edge of the form • $\rightarrow$ •, and, likewise, it contains no two distinct unlabeled vertices whose collections of edges entering into them come from the same set of places.

The syllogistic Π-board $\mathcal{G}$ in contracted form of the partition Σ_0 defined by (3.1) is reported in Figure 3.3(b). Vertices A, B, and C are the relevant unlabeled vertices of the board (here they have been labeled, just to ease references to them). In particular, vertex A stands for the empty node, vertex B for the node $\{q_0, q_1\}$, and, finally, vertex C for the node $\{q_0, q_2\}$.

Observe that the board $\mathcal{G}$ contains the cycle $q_1 \rightarrow q_2 \rightarrow$ C $\rightarrow q_1$ (corresponding to the cycle $\{q_1\} \rightarrow q_2 \rightarrow \{q_0, q_2\} \rightarrow q_1 \rightarrow \{q_1\}$, expressed in expanded form). The presence of cycles in a syllogistic board is a necessary condition for having infinite formative processes (see Exercise 3.1). When additional constraints that prevent some blocks from becoming infinite are allowed (such as *colors*, in colored boards; see Section 3.2), the mere presence of cycles may fail to be a sufficient condition for having infinite formative processes. Such a problem will be studied in depth in Chapter 5.   ∎

3.1.3   Special Events and Shadow Processes

In connection with the satisfiability problem for MLSP-like theories, certain events in the formative processes are particularly relevant, and for this reason will be called *special events*. More specifically, for a formative process relative to a $\mathcal{P}$-partition Σ (inducing a syllogistic Π-board $\mathcal{G}$), special events relevant to our purposes are of two types:

1. the ordinals at which some edge $B \rightarrow q$ in $\mathcal{G}$ gets activated for the first time (*edge-activation events*);

2. the triples $\langle \alpha, B, q \rangle$ such that at step α some set of the form $\bigcup \Gamma$, with $\Gamma \subseteq \Sigma$, flows along the edge $B \rightarrow q$ of $\mathcal{G}$ (*grand-events*).

With the aid of Figure 3.4, we can easily determine the edge-activation events and the grand-events in the case of the formative process $(\Pi_\mu)_{\mu \leqslant 6}$ for the partition Σ_0 defined by (3.1). These are reported in Table 3.1.

The importance of special events stems from the definition of *shadow processes* and their basic property.

Let $\widehat{\mathcal{H}}$ and $\mathcal{H}$ be two formative processes sharing the same syllogist board. If the sequence of special events of $\widehat{\mathcal{H}}$ is synchronized with the sequence of special events of $\mathcal{H}$ (in the sense that the two sequences can be put into a one-to-one order-preserving correspondence), then we say that $\widehat{\mathcal{H}}$ is a shadow process of $\mathcal{H}$. The main property of shadow processes is proved in the Shadow Theorem (Section 3.4):

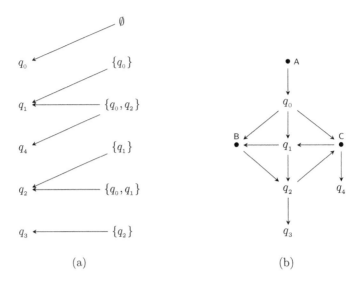

Figure 3.3: Syllogistic Π-board $\mathcal{G}$ of the partition Σ_0, with $\Pi = \{q_0, \ldots, q_4\}$: (a) expanded form; (b) contracted form.

If $\widehat{\mathcal{H}}$ is a shadow process of $\mathcal{H}$, then the final partition of $\widehat{\mathcal{H}}$ imitates that of $\mathcal{H}$ (and, therefore, their final partitions satisfy the same MLSP-formulae).

ν	edges activated	grand-events	trace
0	$\emptyset \to q_0$	$\emptyset \to q_0$	$\emptyset$
1	$\{q_0\} \to q_1$		$\{q_0\}$
2	$\{q_1\} \to q_2$		$\{q_1\}$
3	$\{q_0, q_1\} \to q_2$		$\{q_0, q_1\}$
4	$\{q_2\} \to q_3$	$\{q_2\} \to q_3$	$\{q_2\}$
5	$\{q_0, q_2\} \to q_1$		$\{q_0, q_2\}$
	$\{q_0, q_2\} \to q_4$		

Table 3.1: Edge-activation events, grand-events, and trace of the formative process $\left(\Pi_\mu\right)_{\mu \leqslant 6}$ of Σ_0.

3.1.4 Applications to the Decision Problem of MLSP-like Theories

Let $\mathcal{H}$ be a formative process for a $\mathcal{P}$-partition Σ, and $\mathfrak{S}_{\mathcal{H}}$ the sequence of its flow-distribution steps. Using the techniques first presented in [Can91] and further elaborated in [COU02], it can be proved that it is possible to 'extract' from $\mathfrak{S}_{\mathcal{H}}$ a subsequence $\mathfrak{S}'$,

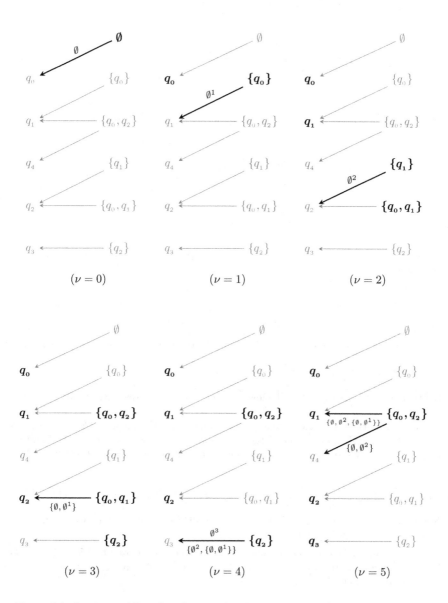

Figure 3.4: Sequence of flow distributions for the construction of the partition Σ_0.

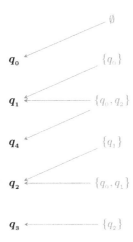

Figure 3.5: Final configuration of the syllogistic Π-board induced by Σ_0: all nodes are inactive.

whose length is bounded by the value $c(|\Sigma|)$ for a suitable computable function $c(\cdot)$, such that

1. $\mathfrak{S}'$ generates a formative process $\widehat{\mathcal{H}}$ for a $\mathcal{P}$-partition $\widehat{\Sigma}$ such that $\mathsf{rk}\,\widehat{\Sigma} \leqslant c(|\Sigma|)$, and

2. $\widehat{\mathcal{H}}$ is a shadow process of $\mathcal{H}$.

Hence, as a consequence of the Shadow Theorem and Corollaries 2.31 and 2.25, it follows that $\widehat{\Sigma}$ satisfies all normalized MLSP-conjunctions satisfied by Σ.

The above argument yields a *small model property* for the theory MLSP. Indeed, given a normalized MLSP-conjunction Φ, then, as we already know, Φ is satisfiable if and only if it is satisfied by some $\mathcal{P}$-partition Σ whose cardinality does not exceed $2^{|\mathrm{Vars}(\Phi)|}$. Hence, it follows that:

Φ is satisfiable if and only if it is satisfied by a $\mathcal{P}$-partition $\widehat{\Sigma}$, with $\mathsf{rk}\,\widehat{\Sigma} \leqslant c(2^{|\mathrm{Vars}(\Phi)|})$.

Notice that, as a by-product, we obtain that an MLSP-formula is satisfiable if and only if it is hereditarily finitely satisfiable.

Matters become much more complicated when Φ is a normalized MLSSPF-conjunction, i.e., if, in addition to the MLSP-constructs, the conjunction Φ may also contain occurrences of literals of the following types:

$$\{x_1, \ldots, x_n\}, \qquad Finite(x), \qquad \neg Finite(x).$$

Let Φ^- be the conjunction of the literals present in Φ, save those involving the finiteness predicate $Finite(\cdot)$. As will be shown in Chapter 5 (see also [CU14]), if Φ is satisfiable, then the reduced conjunction Φ^- is satisfied by a "small" $\mathcal{P}$-partition $\widehat{\Sigma}$, which, roughly speaking, satisfies the following additional condition:

> The syllogistic board induced by $\widehat{\Sigma}$ contains suitable cycles whose vertices are collectively active at the same time, relative to a formative process for $\widehat{\Sigma}$. Flow-distribution steps can be executed infinite times along such cycles (*pumping mechanism*), yielding a new formative process whose final partition $\widehat{\Sigma}_\infty$ satisfies the initial MLSSPF-conjunction Φ.

Then, the decidability of the satisfiability problem for MLSSPF-formulae readily follows.

We illustrate the pumping mechanism indicated in the above condition with the help of the syllogistic Π-board $\mathcal{G}$ in Figure 3.3(a) for the partition Σ_0 defined by (3.1). To this end, it is more convenient to refer to the contracted version of $\mathcal{G}$ reported in Figure 3.3(b), where the cycle $q_1 \to q_2 \to C \to q_1$ (corresponding, as remarked earlier, to the cycle $\{q_1\} \to q_2 \to \{q_0, q_2\} \to q_1 \to \{q_1\}$ expressed in expanded form) can be easily identified. Thus, the trace

$$\emptyset, \ \{q_0\}, \ \{q_1\}, \ \{q_0, q_1\}, \ \{q_2\}, \ \{q_0, q_2\}$$

of the formative process $(\Pi_\mu)_{\mu \leqslant 6}$ for our partition Σ_0 can be formally extended by inserting in it α iterations of the cycle $(\{q_0, q_2\}, \{q_1\})$, for any ordinal α, yielding the sequence[2]

$$\emptyset, \ \{q_0\}, \ \{q_1\}, \ \{q_0, q_1\}, \ \{q_2\}, \ \left(\{q_1\}, \ \{q_0, q_2\}\right)^\alpha, \ \{q_0, q_2\} \tag{3.3}$$

(here we are borrowing the power notation from the theory of formal languages, with the obvious meaning).

In the specific case under consideration, it can be shown that, for every ordinal α, there exists a formative process having the sequence (3.3) as a trace, and which, additionally, is a shadow process of $(\Pi_\mu)_{\mu \leqslant 6}$.

We limit ourselves to give some hints just for the case $\alpha = \omega$.

Define recursively the sequence of sets $(a_n)_{n < \omega}$, by putting

$$\begin{cases} a_0 &= \emptyset^2 \\ a_{n+1} &= \{\{\emptyset^1, a_n\}\}, \quad \text{for } n \in \omega, \end{cases}$$

and let

$$A_1 := \{\{\emptyset, a_0\}, \{\emptyset, a_1\}, \dots\}$$
$$A_2 := \{a_0, a_1, \dots\}.$$

Further, set

$$\Sigma_\infty := \{q_0^{[\bullet]}, q_1^{[\bullet]}, \dots, q_4^{[\bullet]}\}, \tag{3.4}$$

where

$$q_0^{[\bullet]} = \{\emptyset\}, \qquad q_1^{[\bullet]} = A_1 \cup \{\emptyset^1\}, \qquad q_2^{[\bullet]} = A_2 \cup \{\{\emptyset, \emptyset^1\}\},$$
$$q_3^{[\bullet]} = \{\emptyset^3, q_2^{[\bullet]}\}, \qquad q_4^{[\bullet]} = \{\{\emptyset, \emptyset^2\}\}.$$

---

[2]However, one has to bear in mind that a sequence of nodes obtained by the above pumping mechanism does not have necessarily to be a valid trace of some formative process, unless suitable conditions are satisfied.

Then, it can be shown that the transitive partition Σ_∞ imitates Σ_0. In addition, there is a formative process for Σ_∞, which is a shadow process for $(\Pi_\mu)_{\mu \leq 6}$ and whose trace is

$$\emptyset, \ \{q_0\}, \ \{q_1\}, \ \{q_0, q_1\}, \ \{q_2\}, \ \Big(\{q_1\}, \ \{q_0, q_2\}\Big)^\omega, \ \{q_0, q_2\}. \tag{3.5}$$

3.2   Basics on Syllogistic Boards

We consider a *finite* set Π, whose elements are called PLACES (or SYNTACTICAL VENN REGIONS) and whose subsets are called NODES.[3] We assume that $\Pi \cap \mathcal{P}(\Pi) = \emptyset$, so that no node is a place, and vice versa. We shall use these places and nodes as the vertices of a directed bipartite graph $\mathcal{G}$ of a special kind, called (SYLLOGISTIC) Π-BOARD.[4]

The edges issuing from each place q are exactly all pairs $\langle q, B \rangle$ such that $q \in B \subseteq \Pi$: these are called MEMBERSHIP EDGES. The remaining edges of $\mathcal{G}$, called DISTRIBUTION EDGES, go from nodes to places; hence, $\mathcal{G}$ is fully characterized by the function

$$\mathcal{T} \in \mathcal{P}(\Pi)^{\mathcal{P}(\Pi)}$$

associating with each node B the set of all places t such that $\langle B, t \rangle$ is an edge of $\mathcal{G}$. The elements of $\mathcal{T}(B)$ are the TARGETS of B, and $\mathcal{T}$ is the TARGET FUNCTION of $\mathcal{G}$. Thus, we usually represent $\mathcal{G}$ by $\mathcal{T}$.

Places and nodes of a Π-board are meant to represent the blocks σ and the subsets Γ (or, quite often, their unionsets $\bigcup \Gamma$), respectively, of a $\mathcal{P}$-partition Σ such that $|\Sigma| = |\Pi|$. Thus, if we indicate with $q^{(\bullet)}$ the block corresponding to the place q, and with $B^{(\bullet)}$ the set $\{q^{(\bullet)} : q \in B\}$ corresponding to the node B, our understanding is that

(i) $q^{(\bullet)} \neq \emptyset$, for all $q \in \Pi$;

(ii) $q^{(\bullet)} \cap p^{(\bullet)} = \emptyset$ when $q \neq p$, for all $q, p \in \Pi$;

(iii) there is a place $\bar{q} \in \Pi$ such that

$$\bigcup(\Pi \setminus \{\bar{q}\})^{(\bullet)} \subseteq \bigcup\bigcup \Pi^{(\bullet)} \subseteq \mathcal{P}\Big(\bigcup\bigcup \Pi^{(\bullet)}\Big) = \bigcup \Pi^{(\bullet)}.$$

Notice that (i)–(iii) imply that $\Pi^{(\bullet)}$ is indeed a $\mathcal{P}$-partition.

The intended meaning of $t \in \mathcal{T}(B)$ is that $t^{(\bullet)}$ intersects the set $\mathcal{P}^*(B^{(\bullet)})$, that is, in symbols,

$$t \in \mathcal{T}(B) \ \overset{\text{Def}}{:\Longleftrightarrow} \ t^{(\bullet)} \cap \mathcal{P}^*(B^{(\bullet)}) \neq \emptyset, \tag{3.6}$$

where $\mathcal{P}^*(\cdot)$ is the variant of the powerset operator defined in (1.6). More precisely, not only $t^{(\bullet)} \cap \mathcal{P}^*(B^{(\bullet)}) \neq \emptyset$ must hold when $t \in \mathcal{T}(B)$, but also $s^{(\bullet)} \cap \mathcal{P}^*(B^{(\bullet)}) = \emptyset$ when $s \in \Pi \setminus \mathcal{T}(B)$. Thus, $\mathcal{T}(B)$ is the minimal subset $\overline{B}$ of Π such that

$$\mathcal{P}^*(B^{(\bullet)}) \cap \bigcup \Pi^{(\bullet)} \ \subseteq \ \bigcup \overline{B}^{(\bullet)} \tag{3.7}$$

---

[3]Intuitively, places are empty cells, whereas blocks are places filled of sets.

[4]When the set of places Π is understood, we shall also use the expression (SYLLOGISTIC) BOARD.

holds. In other words, for a node B, the members of $\mathcal{P}^*(B^{(\bullet)})$ belonging also to the domain of $\Pi^{(\bullet)}$ are distributed among *all and only* the sets $t^{(\bullet)}$ corresponding exactly to the targets of B. From (3.7), we also obtain the following dual inclusion:

$$t^{(\bullet)} \cap \mathcal{P}(\bigcup \Pi^{(\bullet)}) \subseteq \bigcup_{\substack{B \subseteq \Pi \\ t \in \mathcal{T}(B)}} \mathcal{P}^*(B^{(\bullet)}). \tag{3.8}$$

Indeed, let $s \in t^{(\bullet)} \cap \mathcal{P}(\bigcup \Pi^{(\bullet)})$. Thus, $s \subseteq \bigcup \Pi^{(\bullet)}$. Let D be the minimal subset of Π such that $s \subseteq \bigcup D^{(\bullet)}$. Then $s \in \mathcal{P}^*(D^{(\bullet)}) \cap \bigcup \Pi^{(\bullet)}$, so that, by (3.7), $t^{(\bullet)} \cap t'^{(\bullet)} \neq \emptyset$, for some target t' of B. Then (ii) above implies that $t = t'$, i.e., t is a target of B and therefore $s \in \bigcup_{B \subseteq \Pi \wedge t \in \mathcal{T}(B)} \mathcal{P}^*(B^{(\bullet)})$, proving (3.8).

Summing up, the intended semantics of syllogistic Π-boards is the following.

Definition 3.3 (Compliance with a Π-board). *A $\mathcal{P}$-partition Σ is said to* COMPLY WITH *a syllogistic Π-board $\mathcal{G}$ (and, symmetrically, the syllogistic Π-board $\mathcal{G}$ is said to be* INDUCED BY *the $\mathcal{P}$-partition Σ) via the map $q \mapsto q^{(\bullet)}$, where $|\Sigma| = |\Pi|$ and $q \mapsto q^{(\bullet)}$ belongs to Σ^Π, if*

(a) the map $q \mapsto q^{(\bullet)}$ is bijective, and

(b) the target function $\mathcal{T}$ of $\mathcal{G}$ satisfies (3.6), i.e.,

$$\mathcal{T}(B) = \{q \in \Pi \mid q^{(\bullet)} \cap \mathcal{P}^*(B^{(\bullet)}) \neq \emptyset\}$$

for every $B \subseteq \Pi$. ∎

Syllogistic Π-boards are adequate for the satisfiability problem for MLSSP-formulae, as we shall see in the next chapter. To handle the finiteness predicate *Finite*$(\cdot)$ in MLSSPF-formulae, it turns out that syllogistic Π-boards need to be extended by *colors*. Formally, a COLORED Π-BOARD $\mathcal{G} = \langle \mathcal{T}, \mathcal{R} \rangle$ is characterized by

- a target function $\mathcal{T}$, and

- a designated set $\mathcal{R}$ of places.

The places in $\mathcal{R}$ are said to be RED, whereas those in $\Pi \setminus \mathcal{R}$ are GREEN. A node is red if the places in it are all red; otherwise (i.e., if some place in the node is green), it is green. Instead, a list of vertices of $\mathcal{G}$ is green if all its vertices are green.

Informally, a place is red when its corresponding blocks must be finite. Likewise, a node is red if the union of the blocks corresponding to its places must be finite. More precisely, the intended semantics of colored Π-boards is given in the following definition.

Definition 3.4 (Compliance with a colored Π-board). *A $\mathcal{P}$-partition Σ is said to* COM-PLY WITH *a colored Π-board $\mathcal{G} = \langle \mathcal{T}, \mathcal{R} \rangle$ (and, symmetrically, the colored Π-board $\mathcal{G}$ is said to be* INDUCED BY *the $\mathcal{P}$-partition Σ) via the map $q \mapsto q^{(\bullet)}$, where $|\Sigma| = |\Pi|$ and $q \mapsto q^{(\bullet)}$ belongs to Σ^Π, if Σ complies with the uncolored part of $\mathcal{G}$, namely,*

(a) the map $q \mapsto q^{(\bullet)}$ is bijective,

(b) the target function $\mathcal{T}$ of $\mathcal{G}$ satisfies (3.6), i.e.,

$$\mathcal{T}(B) = \{q \in \Pi \mid q^{(\bullet)} \cap \mathcal{P}^*(B^{(\bullet)}) \neq \emptyset\}$$

for every $B \subseteq \Pi$, and, additionally,

(c) $|r^{(\bullet)}| < \aleph_0$, for every $r \in \mathcal{R}$.

The place $\overline{q} \in \Pi$ such that $\overline{q}^{(\bullet)}$ is the special block of Σ is called SPECIAL. *Any node $B \subseteq \Pi$ not containing the special place $\overline{q}$ is called a* $\mathcal{P}$-NODE. ∎

When a (colored) Π-board $\mathcal{G}$ is induced by a $\mathcal{P}$-partition Σ, we shall also refer to it as a Σ-BOARD, and denote it by the pair $\langle \Sigma, \mathcal{G} \rangle$.

Definition 3.5 (Canonical Σ-board). *Let Σ be a $\mathcal{P}$-partition. A Σ-board $\mathcal{G} = \langle \mathcal{T}, \mathcal{R} \rangle$, complying with Σ via the bijection $q \mapsto q^{(\bullet)}$ and whose set of places is Π, is said to be* CANONICAL *if*

$$\mathcal{T} = \{\langle B, q \rangle \mid B \subseteq \Pi, \ q \in \Pi, \ and \ \mathcal{P}^*(B^{(\bullet)}) \cap q^{(\bullet)} \neq \emptyset\}$$
$$\mathcal{R} = \{r \in \Pi \mid Finite(r^{(\bullet)})\}.$$

Let $\Pi = \{q_1, \ldots, q_n\}$, Σ a $\mathcal{P}$-partition with special block $\overline{\sigma}$, and $q \mapsto q^{(\bullet)}$ a bijection in Σ^Π. Also, let us assume, without loss of generality, that $q_n^{(\bullet)} = \overline{\sigma}$ (i.e., q_n is the special place of Π). If a colored Π-board $\mathcal{G} = \langle \mathcal{T}, \mathcal{R} \rangle$ is induced by Σ via the map $q \mapsto q^{(\bullet)}$, then the following constraints on the $q_i^{(\bullet)}$'s hold:

$$\bigwedge_{1 \leqslant i < j \leqslant n} q_i^{(\bullet)} \cap q_j^{(\bullet)} = \emptyset \ \wedge \ \bigwedge_{1 \leqslant j \leqslant n} q_j^{(\bullet)} \neq \emptyset \tag{3.9}$$

$$\bigcup (\Pi \setminus \{q_n\})^{(\bullet)} \subseteq \bigcup\bigcup \Pi^{(\bullet)} \subseteq \mathcal{P}(\bigcup\bigcup \Pi^{(\bullet)}) = \bigcup \Pi^{(\bullet)} \tag{3.10}$$

$$\bigwedge_{\langle B, q \rangle \in \mathcal{T}} \mathcal{P}^*(B^{(\bullet)}) \cap q^{(\bullet)} \neq \emptyset \ \wedge \ \bigwedge_{\langle B, q \rangle \notin \mathcal{T}} \mathcal{P}^*(B^{(\bullet)}) \cap q^{(\bullet)} = \emptyset \tag{3.11}$$

$$\bigwedge_{r \in \mathcal{R}} Finite(r^{(\bullet)}). \tag{3.12}$$

Indeed, the conjunction (3.9) plainly states that the $q_i^{(\bullet)}$'s are non-null and pairwise disjoint (therefore, pairwise distinct), so that the set $\{q_1^{(\bullet)}, \ldots, q_n^{(\bullet)}\}$ is a partition with exactly n blocks (namely, the partition Σ). In addition, the constraint (3.10), together with (3.9), expresses that the partition $\{q_1^{(\bullet)}, \ldots, q_n^{(\bullet)}\}$ is actually a $\mathcal{P}$-partition. Next, the conjunction (3.11), together with (3.9) and (3.10), expresses that the uncolored part of the Π-board $\mathcal{G}$ complies with $\Pi^{(\bullet)} = \Sigma$. Finally, the conjunction (3.12), together with (3.9), (3.10), and (3.11), expresses that the colored Π-board $\mathcal{G}$ complies with $\Pi^{(\bullet)} = \Sigma$.

If the sets $q_i^{(\bullet)}$'s are temporarily regarded as set variables, the conjunctions (3.9) and (3.12) plainly belong to the MLSSPF fragment. Concerning (3.10), the terms $\bigcup (\Pi \setminus \{q_n\})^{(\bullet)}$ and $\bigcup \Pi^{(\bullet)}$ in it can be readily expressed by MLSSPF-terms, as

$$\bigcup (\Pi \setminus \{q_n\})^{(\bullet)} = q_1^{(\bullet)} \cup \ldots \cup q_{n-1}^{(\bullet)} \quad and \quad \bigcup \Pi^{(\bullet)} = q_1^{(\bullet)} \cup \ldots \cup q_n^{(\bullet)}.$$

Unfortunately, this is not the case, in general, for the term $\bigcup\bigcup\Pi^{(\bullet)} = \bigcup\left(q_1^{(\bullet)} \cup \ldots \cup q_n^{(\bullet)}\right)$, since the unary union is not expressible in the theory MLSSPF. However, by exploiting the injectivity of the powerset operator, the constraint $\mathcal{P}(\bigcup\bigcup\Pi^{(\bullet)}) = \bigcup\Pi^{(\bullet)}$ in (3.10) characterizes univocally the term $\bigcup\bigcup\Pi^{(\bullet)}$. Therefore the term $\bigcup\bigcup\Pi^{(\bullet)}$ can just be replaced by a fresh set variable in the corresponding MLSSPF-formula. Finally, the constraints $\mathcal{P}^*(B^{(\bullet)}) \cap q^{(\bullet)} \neq \emptyset$ and $\mathcal{P}^*(B^{(\bullet)}) \cap q^{(\bullet)} = \emptyset$ in (3.11) can be written in terms of the powerset operator, by observing that we have

$$\mathcal{P}^*(B^{(\bullet)}) = \mathcal{P}(\bigcup B^{(\bullet)}) \setminus \bigcup\{\mathcal{P}(\bigcup(B \setminus \{q\})^{(\bullet)}) \mid q \in B\}$$
$$= \mathcal{P}(\bigcup B^{(\bullet)}) \setminus \left(\bigcup_{q \in B} \mathcal{P}(\bigcup(B \setminus \{q\})^{(\bullet)})\right),$$

and that $\mathcal{P}(\bigcup B^{(\bullet)}) \setminus \left(\bigcup_{q \in B} \mathcal{P}(\bigcup(B \setminus \{q\})^{(\bullet)})\right)$ can be written as an MLSSPF-term, for $B \subseteq \Pi$ and $q \in \Pi$. For instance, if $B = \{p, q, r\}$, then we have

$$\mathcal{P}^*(B^{(\bullet)}) = \mathcal{P}(p^{(\bullet)} \cup q^{(\bullet)} \cup r^{(\bullet)}) \setminus \left(\mathcal{P}(q^{(\bullet)} \cup r^{(\bullet)}) \cup \mathcal{P}(p^{(\bullet)} \cup r^{(\bullet)}) \cup \mathcal{P}(p^{(\bullet)} \cup q^{(\bullet)})\right).$$

In conclusion, if $\Phi_{\mathcal{G}}$ is the conjunction of the following MLSSPF-constraints

$$\bigwedge_{1 \leq i < j \leq n} q_i \cap q_j = \emptyset \wedge \bigwedge_{1 \leq j \leq n} q_j \neq \emptyset \wedge \bigwedge_{r \in \mathcal{R}} Finite(r)$$
$$\bigcup(\Pi \setminus \{q_n\}) \subseteq x \subseteq \mathcal{P}(x) = \bigcup\Pi$$
$$\bigwedge_{\langle B, q\rangle \in \mathcal{T}} \left(\mathcal{P}(\bigcup B) \setminus \left(\bigcup_{q \in B} \mathcal{P}(\bigcup(B \setminus \{q\}))\right)\right) \cap q \neq \emptyset$$
$$\bigwedge_{\langle B, q\rangle \notin \mathcal{T}} \left(\mathcal{P}(\bigcup B) \setminus \left(\bigcup_{q \in B} \mathcal{P}(\bigcup(B \setminus \{q\}))\right)\right) \cap q = \emptyset,$$

where the places q_i's are regarded as set variables (namely, names of unknown sets) and x is a set variable distinct from the q_i's, then $\Phi_{\mathcal{G}}$ is satisfiable if and only if the colored Π-board $\mathcal{G}$ is induced by some $\mathcal{P}$-partition of size n. Thus, we may view a colored board $\mathcal{G} = \langle \mathcal{T}, \mathcal{R}\rangle$ as being the representation of a suitable conjunction $\Phi_{\mathcal{G}}$ of constraints belonging to MLSSPF. When each 'variable' q in $\Phi_{\mathcal{G}}$ is replaced by a set $q^{(\bullet)}$ in such a way that the overall substitution satisfies $\Phi_{\mathcal{G}}$, then the sets $q^{(\bullet)}$ will collectively form a $\mathcal{P}$-partition Σ actually complying with $\mathcal{G}$.

Let $\widehat{\Phi}_{\mathcal{G}}$ be the conjunction obtained by adding to $\Phi_{\mathcal{G}}$ all constraints $\neg Finite(q)$, for $q \in \Pi \setminus \mathcal{R}$. By investigating how we can get a solution to $\widehat{\Phi}_{\mathcal{G}}$ out of a solution to Φ (if possible), we shall shed light on the conditions for the satisfiability of any formula of the quantifier-free theory MLSSPF; and, ultimately, we shall produce an algorithm to test the satisfiability of any MLSSPF-formula.

Let Φ be a normalized MLSSPF-conjunction, and M a model satisfying Φ. If we color a $\Sigma_M^{\mathcal{P}}$-board $\mathcal{G}$ induced by M by letting $\mathcal{R}$ be the collection of places q of $\mathcal{G}$ such that, for some set variable $x \in \mathrm{Vars}(\Phi)$, we have

(i) $q^{(\bullet)} \subseteq Mx$, and

(ii) the conjunction Φ contains either the literal $Finite(x)$ or a literal of the form $x = \{y_1, \ldots, y_H\}$, then we obtain the so-called CANONICAL COLORED BOARD INDUCED BY M AND Φ (in short, a $\Sigma_{M,\Phi}$-board).

3.3   Formative Processes and Basic Events

We shall discuss in detail a technique by which one can find a $\mathcal{P}$-partition

$$\Sigma_\xi = \{q^{(\bullet)} : q \in \Pi\}$$

complying with a given Π-board by

- constructing a ξ-sequence of partitions, which is a *formative process* in the sense specified by Definition 3.6 below,

- ascertaining that every edge $\langle A, t\rangle$ of the Π-board has been *activated* along the process (in a sense to be clarified below), and

- taking as Σ_ξ the last partition of the sequence.

Definition 3.6 (Formative process, trace, and history). *Let Σ, Σ' be partitions such that*

- *every block $\sigma \in \Sigma$ has a block $\sigma' \in \Sigma'$ for which $\sigma \subseteq \sigma'$, and*

- *$\sigma_0, \sigma_1 \subseteq \tau$ implies $\sigma_0 = \sigma_1$, when $\sigma_0, \sigma_1 \in \Sigma$ and $\tau \in \Sigma'$.*

In this case, we say that Σ' FRAMES Σ (or, equivalently, that Σ' ENLARGES Σ).

   Whenever

- *Σ' frames Σ,*

- *$\Sigma \neq \Sigma'$, and*

- *$\bigcup \Sigma' \setminus \bigcup \Sigma \subseteq \mathcal{P}^*(\Gamma)$, for some $\Gamma \subseteq \Sigma$,*

the ordered pair $\langle \Sigma, \Sigma'\rangle$ is called an ACTION (VIA Γ) (or a PROLONGATION).

   A (FORMATIVE) PROCESS is a sequence $\left(\Pi_\mu\right)_{\mu \leqslant \xi}$ of partitions, where the LENGTH ξ of the process can be any ordinal, and, for every ordinal $\nu < \xi$ and every limit ordinal λ such that $\nu < \lambda \leqslant \xi$,

- *$\Pi_\nu = \emptyset$, if $\nu = 0$;*

- *$\langle \Pi_\nu, \Pi_{\nu+1}\rangle$ is an action;*

- *Π_λ frames Π_ν,[5] and*

$$\bigcup \Pi_\lambda \subseteq \bigcup \left\{\bigcup \Pi_\gamma : \gamma < \lambda\right\}.$$

---

[5]More in general, Π_μ frames Π_ν, for $0 \leqslant \nu < \mu \leqslant \xi$; cf. Exercise 3.4.

For all $\mu \leqslant \xi$ and $\tau \in \Pi_\xi$, we put

$$\tau^{(\mu)} := \begin{cases} \text{the unique } \sigma \in \Pi_\mu \text{ such that } \sigma \subseteq \tau, & \text{if any exists} \\ \emptyset & \text{otherwise.} \end{cases} \quad (3.13)$$

A process $\left(\Pi_\mu\right)_{\mu \leqslant \xi}$ is said to be GREEDY *(or* STRONG*) if, for all $\nu < \xi$ and $\Gamma \subseteq \Pi_\nu$:*[6]

$$\mathcal{P}^*(\Gamma) \cap \left(\bigcup \Pi_{\nu+1} \setminus \bigcup \Pi_\nu\right) \neq \emptyset \quad \text{implies} \quad \left(\mathcal{P}^*(\Gamma) \setminus \bigcup \Pi_{\nu+1}\right) \cap \bigcup \Pi_\xi = \emptyset.$$

To stress the fact that no greediness assumption is made for a given formative process, this may be referred to as a WEAK *process.*[7]

A formative process $\left(\Pi_\mu\right)_{\mu \leqslant \xi}$, whose final partition Π_ξ is a $\mathcal{P}$-partition complying with a Π-board $\mathcal{T}$ via a bijection $q \mapsto q^{(\bullet)}$ in $(\Pi_\xi)^\Pi$, is called a Π-PROCESS, *and is represented by the triple $\left\langle \left(\Pi_\mu\right)_{\mu \leqslant \xi}, (\bullet), \mathcal{T} \right\rangle$.*

Similarly, a formative process $\left(\Pi_\mu\right)_{\mu \leqslant \xi}$, whose final partition Π_ξ is a $\mathcal{P}$-partition complying with a colored Π-board $\langle \mathcal{T}, \mathcal{R} \rangle$ via a bijection $q \mapsto q^{(\bullet)}$ in $(\Pi_\xi)^\Pi$, is called a COLORED Π-PROCESS, *and is represented by the quadruple $\left\langle \left(\Pi_\mu\right)_{\mu \leqslant \xi}, (\bullet), \mathcal{T}, \mathcal{R} \right\rangle$.*

For a Π-process $\mathcal{H} := \left\langle \left(\Pi_\mu\right)_{\mu \leqslant \xi}, (\bullet), \mathcal{T} \right\rangle$ (or a colored Π-process $\left\langle \left(\Pi_\mu\right)_{\mu \leqslant \xi}, (\bullet), \mathcal{T}, \mathcal{R} \right\rangle$) and, for all $\mu \leqslant \xi$, $\nu < \xi$, $p \in \Pi$, $B \subseteq \Pi$, we designate by $p^{(\mu)}, B^{(\mu)}, B^{(\bullet)}, \Delta^{(\nu)}(p), A_\nu, T_\nu$, respectively, the unique sets such that

$$
\begin{aligned}
p^{(\mu)} &:= \left(p^{(\bullet)}\right)^{(\mu)} \\
B^{(\mu)} &:= \left\{q^{(\mu)} : q \in B\right\} \\
B^{(\bullet)} &:= \left\{q^{(\bullet)} : q \in B\right\} \\
\Delta^{(\nu)}(p) &:= p^{(\nu+1)} \setminus \bigcup \Pi_\nu \\
A_\nu &:= \text{the set } A \subseteq \Pi \text{ for which } \bigcup \Pi_{\nu+1} \setminus \bigcup \Pi_\nu \subseteq \mathcal{P}^*\left(A^{(\nu)}\right) \\
T_\nu &:= \left\{p \in \Pi \mid \Delta^{(\nu)}(p) \cap \left(\bigcup \Pi_{\nu+1} \setminus \bigcup \Pi_\nu\right) \neq \emptyset\right\}.
\end{aligned}
$$

We shall call

- *A_ν, the ν-TH* MOVE *of the process,*

- *$(A_\nu)_{\nu < \xi}$, the* TRACE *of the process, and*

- *$(A_\nu, T_\nu)_{\nu < \xi}$, the* HISTORY *of the process (or, sometimes, the* HISTORY *of the final $\mathcal{P}$-partition Π_ξ).* ∎

Remark 3.7. In some situations, we shall deal with two Π-processes, $\mathcal{H}$ and $\widehat{\mathcal{H}}$, complying with the same Π-board $\langle \mathcal{T} \rangle$ (or colored Π-board $\langle \mathcal{T}, \mathcal{R} \rangle$). For this purpose, we introduce a variation of the notations

$$\left\langle \left(\Pi_\mu\right)_{\mu \leqslant \xi}, (\bullet), \mathcal{T} \right\rangle, \quad p^{(\mu)}, \quad B^{(\mu)}, \quad B^{(\bullet)}, \quad \Delta^{(\nu)}(p), \quad A_\nu$$

---

[6]The greediness condition is also referred to as COHERENCE REQUIREMENT.

[7]Though every weak process can be strenghtened to a greedy process (cf. [COU02, Section 6]), shadow processes, which will play an important role in the second part of the book, arise naturally as weak processes (cf. Section 5.2).

introduced for $\mathcal{H}$ at the end of the previous definition.

Specifically, we shall denote the second Π-process $\widehat{\mathcal{H}}$ by $\langle (\widehat{\Pi}_\alpha)_{\alpha \leqslant \widehat{\xi}}, [\bullet], \mathcal{T} \rangle$, where $\widehat{\Pi}_{\widehat{\xi}} = \{q^{[\bullet]} \mid q \in \Pi\}$, and, for all $\alpha \leqslant \widehat{\xi}$, $\beta < \widehat{\xi}$, $p \in \Pi$, $B \subseteq \Pi$, we shall designate by $p^{[\alpha]}, B^{[\alpha]}, B^{[\bullet]}, \Delta^{[\beta]}(p), \widehat{A}_\beta, \widehat{T}_\beta$, respectively, the unique sets such that

$$
\begin{aligned}
p^{[\alpha]} &:= \left(p^{[\bullet]}\right)^{(\alpha)} \\
B^{[\alpha]} &:= \{q^{[\alpha]} : q \in B\} \\
B^{[\bullet]} &:= \{q^{[\bullet]} : q \in B\} \\
\Delta^{[\beta]}(p) &:= p^{[\beta+1]} \setminus \bigcup \widehat{\Pi}_\beta \\
\widehat{A}_\beta &:= \text{the set } A \subseteq \Pi \text{ for which } \bigcup\widehat{\Pi}_{\beta+1} \setminus \bigcup\widehat{\Pi}_\beta \subseteq \mathcal{P}^*\left(A^{[\beta]}\right) \\
\widehat{T}_\beta &:= \{p \in \Pi \mid \Delta^{[\beta]}(p) \cap (\bigcup\widehat{\Pi}_{\beta+1} \setminus \bigcup\widehat{\Pi}_\beta) \neq \emptyset\}. \qquad \blacksquare
\end{aligned}
$$

Lemma 3.8. *Every constituent Π_μ of a formative process $(\Pi_\mu)_{\mu \leqslant \xi}$ is a transitive partition.*

Proof. Let Π_μ, with $\mu \leqslant \xi$, be a constituent of a given formative process $(\Pi_\mu)_{\mu \leqslant \xi}$. Since, by the very definition of formative process, Π_μ is a partition, we only need to show that its domain $\bigcup\Pi_\mu$ is transitive. Let $e \in \bigcup\Pi_\mu$, and $\nu_e \leqslant \mu$ the least ordinal ν such that $e \in \bigcup\Pi_\nu$. It is not hard to show that ν_e is a successor ordinal, say, $\nu_e := \mu_e + 1$ (see Exercise 3.5). Then, the set e makes its appearance in the formation process during the execution of the action $\langle \Pi_{\mu_e}, \Pi_{\mu_e+1} \rangle$ via, say, $\Gamma_{\mu_e} \subseteq \Pi_{\mu_e}$. Hence,

$$
e \in \bigcup\Pi_{\mu_e+1} \setminus \bigcup\Pi_{\mu_e} \subseteq \mathcal{P}^*(\Gamma_{\mu_e}) \subseteq \mathcal{P}(\bigcup\Gamma_{\mu_e}) \subseteq \mathcal{P}(\bigcup\Pi_{\mu_e}) \subseteq \mathcal{P}(\bigcup\Pi_\mu)
$$

(since $\bigcup\Pi_{\mu_e} \subseteq \bigcup\Pi_\mu$), and therefore $e \subseteq \bigcup\Pi_\mu$. Thus, by the arbitrariness of $e \in \bigcup\Pi_\mu$, the transitivity of $\bigcup\Pi_\mu$ follows. $\square$

Next, we shall describe a procedure which, given a weak formative process, constructs a greedy formative process, usually shorter, ending in the same partition as the original process.[8]

The transfinitely recursive procedure shown below receives in input the trace $(A_\nu)_{\nu < \xi}$ (relative to a given set of places Π) of the weak formative process $\mathcal{H}$ to be mimicked, along with the ending bijection $q \mapsto q^{(\bullet)}$ of $\mathcal{H}$, and supplies in output the trace of the mimicking process $\mathcal{H}'$, with indication of how the partitioning function of the latter evolves.

> **procedure**   strengthenProcess($(A_\mu)_{\mu < \xi}$, $\{q^{(\bullet)}\}_{q \in \Pi}$);
> $\quad \gamma := \emptyset;$ -- γ assigns to the ν-th move of $\mathcal{H}'$ the position $\gamma(\nu)$
> $\quad\quad\quad\quad\quad$ -- of the move of $\mathcal{H}$ it mimicks
> $\quad \nabla := \emptyset;$ -- ∇ assigns to the ν-th move of $\mathcal{H}'$ its associated partition
> $\quad$ **for** $q \in \Pi$ **do** $\widehat{q} := \emptyset;$ **end for**;   -- start with void blocks
> $\quad$ **notation:** throughout, and for all $B \subseteq \Pi$, $\widehat{B} := \{\widehat{q} \mid q \in B\}$;
> $\quad \mathcal{T} := \{ \langle B, \{q \in \Pi \mid q^{(\bullet)} \cap \mathcal{P}^*(B^{(\bullet)}) \neq \emptyset\} \rangle \mid B \subseteq \Pi \};$
> $\quad\quad\quad$ -- 'targets' for moves: each $\mathcal{T}(B)$ comprises those q
> $\quad\quad$ -- for which $\mathcal{P}^*(\widehat{B})$ will ever intersect $\widehat{q}$
> $\quad$ **for** μ **in** $[0, 1, \ldots, \xi - 1]$ **do**

---

[8] Another procedure of the same kind will be seen in Section 4.2, where the original formative process will, instead, be assumed to be greedy, and the aim will be just to *simulate* the ending partition through the new process: the latter will no longer be guaranteed to be greedy, but its length will be finite even when the original process is infinite.

```
        if ⋃Π̂ = ⋃Π^(•) then quit for-loop; end if;
        A := A_μ;
        S := ⋃Π^(•) ∩ 𝒫*(Â) \ ⋃Π̂;
        if S ≠ ∅ then
            ν      := ⋃_{α∈dom(γ)} (α + 1);
            γ(ν) := μ;
            assert
                (∃ {Δ(p)}_{p∈𝒯(A)}) (
                    isPartition(Δ, S) ∧
                        (∀ q ∈ Π)( S ∩ q^(•) ≠ ∅ → S ∩ q^(•) ⊆ Δ(q) ) );
            let ∇(ν) be one such Δ;
            let the ν-th move consist of the set A paired with this function ∇(ν);
            for q ∈ 𝒯(A) do q̂ := q̂ ∪ Δ(q); end for;
        end if;
    end for;
    ξ' := ⋃_{α∈dom(γ)} (α + 1);
    return
        (A_{γ(α)}, ∇(α))_{α<ξ'};   -- sequence of mimicking moves
end   strengthenProcess;

procedure  isPartition( Δ, S );
    claim
        if (∀ q, r ∈ dom(Δ))( q ≠ r → Δ(q) ∩ Δ(r) = ∅ ),
        then Δ[dom(Δ)] \ {∅} is a partition ;
    return
        ⋃Δ[dom(Δ)] = S ∧ (∀ q, r ∈ dom(Δ))( q ≠ r → Δ(q) ∩ Δ(r) = ∅ );
end   isPartition.
```

The proof that the formative process returned by strengthenProcess really meets the purpose stated above is left to the reader. Thus we have:

Lemma 3.9. *Let* $(A_\mu)_{\mu<\xi}$ *be the trace of a weak formative process for a $\mathcal{P}$-partition* Σ *(relative to some set of places). Then it is possible to extract a subsequence* $(A_{\gamma(\alpha)})_{\alpha<\xi'}$ *from* $(A_\mu)_{\mu<\xi}$, *with* $\xi' \leqslant \xi$ *(so that* $\gamma(\alpha) < \gamma(\beta)$ *for* $\alpha < \beta < \xi'$*), which is the trace of a greedy formative process of* Σ.

Remark 3.10. Formative processes possess some inherent degree of parallelization, as at each move A_ν all the blocks associated with the targets of the node A_ν may receive new elements from $\mathcal{P}^*(A_\nu^{(\nu)})$ at the same time.

To some extent, it is possible to further parallelize formative processes, by allowing parallel moves. To be more precise, given a Π-process $(\Pi_\mu)_{\mu\leqslant\xi}$ with sequence of moves $(A_\nu)_{\nu<\xi}$, for any two ordinals ν, μ such that $\nu < \mu < \xi$, we put $\nu \prec \mu$ whenever A_ν has a target $q \in A_\mu$ that gets new elements from A_ν at step ν, and consider the transitive closure of $\prec$ (which, for convenience, we continue to denote with the same symbol $\prec$). It can be proved that each permutation of the sequence of moves $(A_\nu)_{\nu<\xi}$ that is order-compatible with the strict partial order $\prec$ (including the parallel execution of incomparable moves) is a valid sequence of moves for a Π-process ending with the same final partition Π_ξ of $(\Pi_\mu)_{\mu\leqslant\xi}$. ∎

Through an explicit construction, in Section 3.1 we have shown that the specific transitive partition (3.1) has a greedy formative process. Here we prove that such a result holds for every transitive partition.

Theorem 3.11 (Trace Theorem). *For every transitive partition Σ, there is a greedy formative process for Σ of length ξ such that $|\xi| \leqslant |\bigcup \Sigma|$.*

Proof. Let Σ be a transitive partition. Then, for every ordinal μ:

- if μ is a limit ordinal, we put

$$\Pi_\mu := \bigcup_{\gamma < \mu} \left\{ \bigcup_{\gamma < \nu < \mu} \sigma^{\langle \nu \rangle} \mid \sigma \in \Pi_\gamma \right\},$$

where $\sigma^{\langle \nu \rangle}$ indicates the block τ of Π_ν for which $\sigma \subseteq \tau$;[9] then we have:

$$\bigcup \Pi_\mu \subseteq \bigcup \Sigma. \tag{3.14}$$

- if $\mu = \nu + 1$, for some ordinal ν, and $\Pi_\nu = \Sigma$, we put

$$\Pi_{\nu+1} := \Pi_\nu;$$

- if $\mu = \nu + 1$, for some ordinal ν, and $\Pi_\nu \neq \Sigma$, then we have

$$\bigcup \Pi_\nu \subsetneqq \bigcup \Sigma. \tag{3.15}$$

Let $e \in \bigcup \Sigma \setminus \bigcup \Pi_\nu$ of minimal rank. From the transitivity of Σ and the minimality of $\mathsf{rk}\, e$, it follows that $e \subseteq \bigcup \Pi_\nu$. Hence, there is a $\Gamma_\nu \subseteq \Pi_\nu$ such that $e \in \mathcal{P}^*(\Gamma_\nu)$. Then, setting

$$\Pi_{\nu+1} := \left\{ \sigma \cap \left(\bigcup \Pi_\nu \cup \mathcal{P}^*(\Gamma_\nu) \right) \mid \sigma \in \Sigma \right\} \setminus \{\emptyset\},$$

we have

$$\bigcup \Pi_{\nu+1} \subseteq \bigcup \Sigma. \tag{3.16}$$

Since the sequence of the $\bigcup \Pi_\nu$ strictly increases w.r.t. to the set-inclusion relator $\subseteq$ until it has reached $\bigcup \Sigma$, the inclusions (3.14) and (3.16) imply that there must be an ordinal ξ for which $\bigcup \Pi_\xi = \bigcup \Sigma$ and $|\xi| \leqslant |\bigcup \Sigma|$ and such that $\left(\Pi_\mu \right)_{\mu \leqslant \xi}$ is a greedy formative process of Σ. $\qquad \square$

Corollary 3.12. *If $\langle \mathcal{T}, \mathcal{R} \rangle$ is a colored Π-board induced by a finite $\mathcal{P}$-partition Σ via a bijection $q \mapsto q^{(\bullet)}$ in Σ^Π, then there exists a greedy colored Π-process $\left\langle \left(\Pi_\mu \right)_{\mu \leqslant \xi}, (\bullet), \mathcal{T}, \mathcal{R} \right\rangle$ for Σ such that $|\xi| \leqslant |\bigcup \Sigma|$.*

Example 3.13. Resuming the nesting notation $\emptyset^n$ of Example 1.12, let us take

$$\Sigma = \{p_0^{(\bullet)}, p_1^{(\bullet)}, \ldots, p_5^{(\bullet)}\},$$

where

$p_0^{(\bullet)} = \{\emptyset\},$ $p_1^{(\bullet)} = \{\emptyset^1\},$ $p_2^{(\bullet)} = \{\emptyset^2, \{\emptyset, \emptyset^1\}\},$

$p_3^{(\bullet)} = \{\emptyset^3, \{\{\emptyset, \emptyset^1\}\}\},$ $p_4^{(\bullet)} = \{\{\emptyset, \emptyset^2\}\},$ $p_5^{(\bullet)} = \mathcal{P}\left(\bigcup_{j=0}^4 j^{(\bullet)}\right) \setminus \bigcup_{j=0}^4 p_j^{(\bullet)}.$

---

[9]Notice that, in particular, $\Pi_0 = \emptyset$, as required.

One readily sees that

$$\bigcup\left(\Sigma \setminus \{p_5^{(\bullet)}\}\right) = \bigcup_{j=0}^4 j^{(\bullet)} = \{\emptyset,\,\emptyset^1,\,\emptyset^2,\,\{\emptyset,\emptyset^1\},\,\{\{\emptyset,\emptyset^1\}\},\,\{\emptyset,\emptyset^2\},\,\emptyset^3\},$$

which is a transitive set. Hence, we have:

$$\bigcup \Sigma = \bigcup_{i=0}^5 p_i^{(\bullet)} = \mathcal{P}\left(\bigcup_{j=0}^4 p_j^{(\bullet)}\right),$$

so that $\bigcup\bigcup\Sigma = \bigcup_{j=0}^4 p_j^{(\bullet)}$, and, therefore,

$$\left(\Sigma \setminus \{p_5^{(\bullet)}\}\right) = \bigcup\bigcup\Sigma \subseteq \mathcal{P}(\bigcup\bigcup\Sigma) = \bigcup\Sigma,$$

showing that Σ is a $\mathcal{P}$-partition with $p_5^{(\bullet)}$ as special block.

With the elements s of $\bigcup\left(\Sigma \setminus \{p_5^{(\bullet)}\}\right)$ ordered by non-decreasing ranks, we easily associate with each of them the set A for which $s \in \mathcal{P}^*(\{\,q^{(\bullet)} \mid q \in A\,\})$:

s	$\emptyset$	$\emptyset^1$	$\emptyset^2$	$\{\emptyset,\emptyset^1\}$	$\{\{\emptyset,\emptyset^1\}\}$	$\{\emptyset,\emptyset^2\}$	$\emptyset^3$
A	$\emptyset$	$\{0\}$	$\{1\}$	$\{0,1\}$	$\{2\}$	$\{0,2\}$	$\{2\}$

In this concrete example, the construction of the Trace Theorem proceeds according to the following table:

ν	s	A_ν	T_ν	$p_0^{(\nu)}$	$p_1^{(\nu)}$	$p_2^{(\nu)}$	$p_3^{(\nu)}$	$p_4^{(\nu)}$	$p_5^{(\nu)}$
0	$\emptyset$	$\emptyset$	$\{0\}$	$\emptyset$	$\emptyset$	$\emptyset$	$\emptyset$	$\emptyset$	0
1	$\emptyset^1$	$\{0\}$	$\{1\}$	$\{\boldsymbol{\emptyset}\}$	$\emptyset$	$\emptyset$	$\emptyset$	$\emptyset$	0
2	$\emptyset^2$	$\{1\}$	$\{2\}$	$\{\emptyset\}$	$\{\boldsymbol{\emptyset^1}\}$	$\emptyset$	$\emptyset$	$\emptyset$	0
3	$\{\emptyset,\emptyset^1\}$	$\{0,1\}$	$\{2\}$	$\{\emptyset\}$	$\{\emptyset^1\}$	$\{\boldsymbol{\emptyset^2}\}$	$\emptyset$	$\emptyset$	0
4	$\{\emptyset,\emptyset^2\}$	$\{0,2\}$	$\{4\}$	$\{\emptyset\}$	$\{\emptyset^1\}$	$\{\boldsymbol{\emptyset^2,\{\emptyset,\emptyset^1\}}\}$	$\emptyset$	$\emptyset$	0
5	$\emptyset^3$	$\{2\}$	$\{3,5\}$	$\{\emptyset\}$	$\{\emptyset^1\}$	$\{\emptyset^2,\{\emptyset,\emptyset^1\}\}$	$\emptyset$	$\{\boldsymbol{\{\emptyset,\emptyset^2\}}\}$	2
6	$\cdots$	$\cdots$	$\{5\}$	$\{\emptyset\}$	$\{\emptyset^1\}$	$\{\emptyset^2,\{\emptyset,\emptyset^1\}\}$	$\{\boldsymbol{\emptyset^3,\{\{\emptyset,\emptyset^1\}\}}\}$	$\{\{\emptyset,\emptyset^2\}\}$	3
$\vdots$	$\vdots$	$\vdots$	$\vdots$	$\vdots$	$\vdots$	$\vdots$	$\vdots$	$\vdots$	$\vdots$
31	$\cdots$	$\{0,1,2,3,4\}$	$\{5\}$	$\{\emptyset\}$	$\{\emptyset^1\}$	$\{\emptyset^2,\{\emptyset,\emptyset^1\}\}$	$\{\emptyset^3,\{\{\emptyset,\emptyset^1\}\}\}$	$\{\{\emptyset,\emptyset^2\}\}$	112
32	$-$	$-$	$-$	$\{\emptyset\}$	$\{\emptyset^1\}$	$\{\emptyset^2,\{\emptyset,\emptyset^1\}\}$	$\{\emptyset^3,\{\{\emptyset,\emptyset^1\}\}\}$	$\{\{\emptyset,\emptyset^2\}\}$	121

(For convenience, the table reports only the cardinality of the blocks $p_5^{(\nu)}$, rather than their full extension. In addition, updated blocks have been boldfaced.)

The sequence $(\Pi_\nu)_{\nu \leqslant 32}$, where

$$\Pi_\nu = \{p_i^{(\nu)} \mid 0 \leqslant i \leqslant 5\} \setminus \{\emptyset\}, \quad \text{for } 0 \leqslant \nu \leqslant 32,$$

is a formative process of Σ, with trace $(A_\nu)_{\nu<32}$ and history $(A_\nu, T_\nu)_{\nu<32}$, constructed according to the Trace Theorem.

In particular, we have:

$$\Pi_0 \setminus \{p_5^{(0)}\} = \emptyset$$
$$\Pi_1 \setminus \{p_5^{(1)}\} = \{\{\emptyset\}\}$$
$$\Pi_2 \setminus \{p_5^{(2)}\} = \{\{\emptyset\}, \{\emptyset^1\}\}$$
$$\Pi_3 \setminus \{p_5^{(3)}\} = \{\{\emptyset\}, \{\emptyset^1\}, \{\emptyset^2\}\}$$
$$\Pi_4 \setminus \{p_5^{(4)}\} = \{\{\emptyset\}, \{\emptyset^1\}, \{\emptyset^2, \{\emptyset, \emptyset^1\}\}\}$$
$$\Pi_5 \setminus \{p_5^{(5)}\} = \{\{\emptyset\}, \{\emptyset^1\}, \{\emptyset^2, \{\emptyset, \emptyset^1\}\}, \{\{\emptyset, \emptyset^2\}\}\}$$
$$\Pi_6 \setminus \{p_5^{(6)}\} = \{\{\emptyset\}, \{\emptyset^1\}, \{\emptyset^2, \{\emptyset, \emptyset^1\}\}, \{\emptyset^3, \{\{\emptyset, \emptyset^1\}\}\}, \{\{\emptyset, \emptyset^2\}\}\}$$
$$\Pi_7 \setminus \{p_5^{(7)}\} = \{\{\emptyset\}, \{\emptyset^1\}, \{\emptyset^2, \{\emptyset, \emptyset^1\}\}, \{\emptyset^3, \{\{\emptyset, \emptyset^1\}\}\}, \{\{\emptyset, \emptyset^2\}\}\}$$
$$\vdots \qquad \vdots \qquad \vdots$$
$$\Pi_{32} \setminus \{p_5^{(32)}\} = \{\{\emptyset\}, \{\emptyset^1\}, \{\emptyset^2, \{\emptyset, \emptyset^1\}\}, \{\emptyset^3, \{\{\emptyset, \emptyset^1\}\}\}, \{\{\emptyset, \emptyset^2\}\}\}. \qquad \blacksquare$$

Example 3.14. The charts in Figure 3.3 illustrate the construction of a $\mathcal{P}$-partition

$$\Sigma := \{q_0^{(\bullet)}, q_1^{(\bullet)}, \ldots, q_5^{(\bullet)}\}$$

with six blocks by a formative process $(\Pi_\mu)_{\mu \leqslant \xi}$, where $\Pi := \{q_0, q_1, \ldots, q_5\}$.

The first chart represents the partition Σ itself, whereas the subsequent charts display the stages Π_μ of the formative process, as they evolve from the initial stage Π_0, at which all $q_i^{(0)}$'s are empty, up to the final stage Π_ξ, when $q_i^{(\xi)} = q_i^{(\bullet)}$, for $0 \leqslant i \leqslant 5$.

Greyed areas represent parts of the blocks that have already been filled in. In particular, light-greyed regions contain elements assigned to their respective blocks during the last distribution step (this case may apply only to stages indexed by successor ordinals). For instance, at stage α, the blocks corresponding to the places q_0, q_2, and q_3 are partially filled in (but none of them has been completed yet), whereas those corresponding to the places q_1 and q_4 are still empty. In addition, the blocks corresponding to q_2 and q_3 have received new elements during the distribution step $(\alpha - 1)$ (hence, α must be a successor ordinal).

Stage $(\alpha + 1)$ is the result of the α-th distribution step with some non-null $\Gamma \subseteq \{q_0^{(\alpha)}, q_2^{(\alpha)}, q_3^{(\alpha)}\}$. The light-greyed areas correspond to non-null mutually disjoint subsets of $\mathcal{P}^*(\{q_0^{(\alpha)}, q_2^{(\alpha)}, q_3^{(\alpha)}\})$ assigned to the blocks $q_1^{(\alpha+1)}$, $q_4^{(\alpha+1)}$, and $q_5^{(\alpha+1)}$ during such a step. Notice that, at this stage, all blocks are non-null and block $q_5^{(\alpha+1)}$ has reached its final configuration. $\qquad \blacksquare$

Relative to a colored formative Π-process $(\Pi_\mu)_{\mu \leqslant \xi}$, whose sequence of moves is $(A_\nu)_{\nu < \xi}$, we can identify certain *events*: these, roughly speaking, are ordinals ν at which something important happens. Examples of events are, for any $\Gamma \subseteq \Pi_\xi$ and $\sigma \in \Pi_\xi$:

Edge-activation move: a $\nu < \xi$ such that $\sigma^{(\nu+1)} \cap \mathcal{P}^*(\Gamma) \neq \emptyset$ and $\sigma^{(\nu)} \cap \mathcal{P}^*(\Gamma) = \emptyset$;

Grand-event move: a $\nu < \xi$ such that $\bigcup \Gamma \in \bigcup \Pi_{\nu+1} \setminus \bigcup \Pi_\nu$.

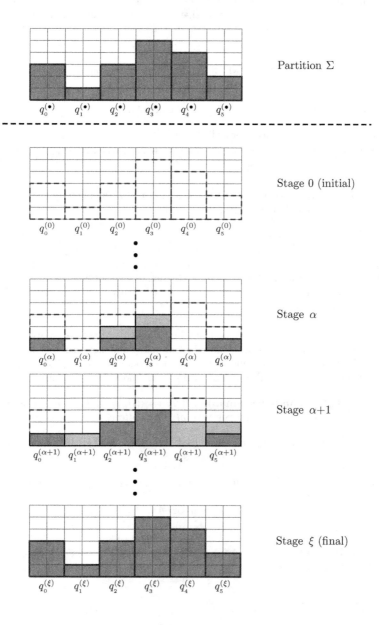

Partition Σ

Stage 0 (initial)

Stage α

Stage $\alpha+1$

Stage ξ (final)

Figure 3.6: Outline of a formative process.

The above events belong to the timeline associated with the sequence of moves $\left(A_\nu\right)_{\nu < \xi}$, and it makes no difference at all whether they occur at a successor or at a limit ordinal.

Other events, instead, more naturally refer to the timeline associated with the formative process $(\Pi_\mu)_{\mu \leqslant \xi}$ itself. This is the case, for instance, for the following event:

Least-Inf event: least $\mu \leqslant \xi$ at which the block $q^{(\mu)}$ corresponding to a place q becomes infinite.

There are two modes by which a block can become infinite:

- *by propagation* at a successor ordinal $\nu + 1$, that is, by drawing infinite elements at step ν from a node B which is already infinite (i.e., when $B = A_\nu$, $\left|\mathcal{P}^*\left(A_\nu^{(\nu)}\right)\right| \geqslant \aleph_0$, and $\left|q^{(\nu)}\right| < \aleph_0 \leqslant \left|q^{(\nu+1)}\right|$); and

- *by core production* at a limit ordinal λ, that is, by drawing finitely many elements at each step of an infinite sequence of moves A_ν, with $\nu < \lambda$ (i.e., when $q^{(\lambda)} = \bigcup_{\nu < \lambda} q^{(\nu)}$ and $\left|q^{(\nu)}\right| < \aleph_0 \leqslant \left|q^{(\lambda)}\right|$, for $\nu < \lambda$).

Plainly, the first ordinal at which some block (associated with a green place) becomes infinite must be ω, and this event can occur by core production only. It is also possible that other blocks become infinite by core production at limit ordinals strictly greater than ω. Later on, we shall identify the conditions that enable such infinite expansions (PUMPING CYCLES; see Section 5.3).

In connection with the events mentioned before, one can define some useful maps. To begin with, recalling the notion of grand-event move and noticing that such an event occurs, in a colored process, at most once for each node, we give the following definition of GRAND-EVENT MAP $\mathsf{GE}(A)$, for A ranging over the nodes of a Π-board.

Definition 3.15 (Grand-event map). *Let* $\mathcal{H} := \left\langle (\Pi_\mu)_{\mu \leqslant \xi}, (\bullet), \mathcal{T}, \mathcal{R} \right\rangle$ *be a colored Π-process. We define the* GRAND-EVENT MAP *of $\mathcal{H}$ by putting, for every node $A \subseteq \Pi$:*

$$\mathsf{GE}(A) := \begin{cases} \text{the ordinal } \nu \text{ for which } \bigcup A^{(\bullet)} \in \bigcup \Pi^{(\nu+1)} \setminus \bigcup \Pi^{(\nu)}, & \text{if any exists,} \\ \text{the length } \xi \text{ of the process,} & \text{otherwise.} \end{cases}$$

Moreover, with a slight abuse of notation, for any given non-null collection $\mathcal{A}$ of nodes, we put

$$\mathsf{GE}(\mathcal{A}) := \min\{\mathsf{GE}(A) : A \in \mathcal{A}\}.$$ ∎

Remark 3.16. Observe that, for every node A of a colored Π-process $\left\langle (\Pi_\mu)_{\mu \leqslant \xi}, (\bullet), \mathcal{T}, \mathcal{R} \right\rangle$, we have

$$\mathsf{GE}(A) = \xi \quad \Longleftrightarrow \quad \bar{q} \in A,$$

where $\bar{q}^{(\bullet)}$ is the special block of Π_ξ, i.e.,

$$\mathsf{GE}(A) = \begin{cases} \text{the ordinal } \nu \text{ for which } \bigcup A^{(\bullet)} \in \bigcup \Pi^{(\nu+1)} \setminus \bigcup \Pi^{(\nu)}, & \text{if } \bar{q} \notin A, \\ \text{the length } \xi \text{ of the process,} & \text{if } \bar{q} \in A. \end{cases}$$ ∎

Likewise, we define the map associated with the place-initialization event, which, for a place of a colored process, returns the largest ordinal at which its corresponding block is still empty.

Definition 3.17 (Edge-activation map). *Let* $\mathcal{H} := \left\langle (\Pi_\mu)_{\mu \leq \xi}, (\bullet), \mathcal{T}, \mathcal{R} \right\rangle$ *be a colored Π-process. We define the* EDGE-ACTIVATION MAP *of $\mathcal{H}$ by putting, for every node $B \subseteq \Pi$ and place $q \in \Pi$:*

$$
\mathsf{EA}(B, q) := \begin{cases} \begin{array}{l} \text{the least ordinal } \nu \text{ such that } B = A_\nu, \\ \quad \mathcal{P}^*\!\left(A_\nu^{(\nu)}\right) \cap q^{(\nu)} = \emptyset, \text{ and } \Delta^{(\nu)}(q) \neq \emptyset, \end{array} & \text{if } \mathcal{P}^*\!\left(B^{(\bullet)}\right) \cap q^{(\bullet)} \neq \emptyset, \\ \text{the length } \xi \text{ of the process,} & \text{otherwise.} \end{cases}
$$

In addition, with a slight abuse of notation, we also put

$$
\mathsf{EA}(q) := \text{the ordinal } \nu \text{ for which } q^{(\nu+1)} \neq \emptyset \text{ and } q^{(\nu)} = \emptyset. \qquad \blacksquare
$$

Finally, we define the map $\mathsf{leastInf}(\cdot)$ associated with the Least-Inf event, which, for a node or place of a colored process, returns the first ordinal of the process at which it becomes infinite.

Definition 3.18 (leastInf map). *Let* $\mathcal{H} := \left\langle (\Pi_\mu)_{\mu \leq \xi}, (\bullet), \mathcal{T}, \mathcal{R} \right\rangle$ *be a colored Π-process. We define the* $\mathsf{leastInf}$ MAP *of $\mathcal{H}$ by putting, for every node $B \subseteq \Pi$:*

$$
\mathsf{leastInf}(B) := \begin{cases} \text{the least ordinal } \mu \text{ for which } \left| \mathcal{P}^*\!\left(B^{(\mu)}\right) \right| \geq \aleph_0, & \text{if } \left| \bigcup B^{(\bullet)} \right| \geq \aleph_0, \\ 0, & \text{otherwise.} \end{cases}
$$

In addition, with a slight abuse of notation, we also put, for every place $q \in \Pi$:

$$
\mathsf{leastInf}(q) := \mathsf{leastInf}(\{q\}). \qquad \blacksquare
$$

Notice that, for a node B and a place q of a given Π-board, $\mathsf{leastInf}(B) \neq 0$ and $\mathsf{leastInf}(t) \neq 0$ imply, respectively, that B and t must be green, otherwise the ending partition would not comply with the color of the board. Notice also that

$$
\mathsf{leastInf}(q) = \begin{cases} \text{the least ordinal } \mu \text{ for which } \left| q^{(\mu)} \right| \geq \aleph_0, & \text{if } \left| q^{(\bullet)} \right| \geq \aleph_0, \\ 0, & \text{otherwise.} \end{cases}
$$

Some simple properties related to the grand-event map are stated in the following lemma, whose proof is left to the reader.

Lemma 3.19. *For every node B of a given colored Π-process $\left\langle (\Pi_\mu)_{\mu \leq \xi}, (\bullet), \mathcal{T}, \mathcal{R} \right\rangle$, we have:*

(a) $B^{(\mathsf{GE}(B))} = B^{(\bullet)}$;

(b) *if* $q^{(\nu+1)} \supsetneq q^{(\nu)}$, *for some $q \in B$ and some $\nu < \xi$, then* $\mathsf{GE}(B) > \nu$;

(c) *if* $q \in B$, *then* $\mathsf{GE}(B) \geq \mathsf{leastInf}(q)$;

(d) *the grand-event map* GE *is injective over the collection of all nodes A such that* $\mathsf{GE}(A) < \xi$;

(e) $\mathsf{GE}(A) > \mathsf{EA}(A)$.

Let A be a node whose grand event takes place at step α. If $\bigcup A^{(\alpha)}$ is infinite, then the α-th move can distribute infinite elements among the targets of A (i.e., the set $\mathcal{P}^*(A^{(\alpha)}) \setminus \bigcup \Pi^{(\alpha)}$ is infinite). This basic fact is proved in the following technical lemma.

Lemma 3.20. *Given a greedy colored* Π-*process* $\langle (\Pi_\mu)_{\mu \leqslant \xi}, (\bullet), \mathcal{T}, \mathcal{R} \rangle$, *consider a node* $A \subseteq \Pi$ *such that* $\bigcup A^{(\bullet)}$ *is infinite and* $\alpha = \mathsf{GE}(A) < \xi$. *Then the set* $\mathcal{P}^*(A^{(\alpha)}) \setminus \bigcup \Pi^{(\alpha)}$ *is infinite.*

Proof. Let $(A_\nu)_{\nu < \xi}$ be the sequence of moves of the given Π-process, and let

$$u_A := \bigcup A^{(\alpha)} \setminus \bigcup A^{(\nu_0)},$$

where $\nu_0 := \min\{\nu \leqslant \alpha \mid A_\nu = A\}$.

If u_A is finite then, by Lemma 1.31(d), the set $S := \{v \in \mathcal{P}^*(A^{(\alpha)}) \mid u_A \subseteq v\}$ is an infinite subset of $\mathcal{P}^*(A^{(\alpha)})$. Thus, it is enough to show that S is disjoint from $\bigcup \Pi^{(\alpha)}$. Assume by contradiction that $S \cap \bigcup \Pi^{(\alpha)} \neq \emptyset$, and let $\bar{v} \in S \cap \bigcup \Pi^{(\alpha)}$. Let ν'_1 be the least ordinal $\nu \leqslant \alpha$ such that $\bar{v} \in \bigcup \Pi^{(\nu)}$. Plainly, ν'_1 is a successor ordinal, so it has the form $\nu'_1 = \nu_1 + 1$. Thus, $A_{\nu_1} = A$ and $u_A \subseteq \bigcup A^{(\nu_1)}$, so that

$$\bigcup A^{(\alpha)} \setminus \bigcup A^{(\nu_0)} \subseteq \bigcup A^{(\nu_1)} \setminus \bigcup A^{(\nu_0)}.$$

This, in turn, implies that $\bigcup A^{(\alpha)} = \bigcup A^{(\nu_1)}$, contradicting the greediness of the process, since $A_{\nu_1} = A = A_\alpha$ and $\nu_1 < \alpha$.

Next, let u_A be infinite. Then the set $\{\bigcup A^{(\alpha)} \setminus \{w\} \mid w \in u_A\}$ is plainly an infinite subset of $\mathcal{P}^*(A^{(\alpha)})$. We claim that

$$\left| \{\bigcup A^{(\alpha)} \setminus \{w\} \mid w \in u_A\} \cap \bigcup \Pi^{(\alpha)} \right| \leqslant 1, \tag{3.17}$$

from which our lemma would readily follow. To prove (3.17), assume by contradiction that there exist distinct $w_1, w_2 \in u_A$ such that

$$\bigcup A^{(\alpha)} \setminus \{w_1\}, \ \bigcup A^{(\alpha)} \setminus \{w_2\} \in \bigcup \Pi^{(\alpha)}.$$

Then there would exist $\nu_1, \nu_2 < \alpha$ such that $A_{\nu_1} = A_{\nu_2} = A$ and $\bigcup A^{(\alpha)} \setminus \{w_i\} \subseteq \bigcup A^{(\nu_i)}$, for $i = 1, 2$. Without loss of generality, suppose $\nu_1 \leqslant \nu_2$. It would then follow that

$$\bigcup A^{(\alpha)} = \left(\bigcup A^{(\alpha)} \setminus \{w_1\} \right) \cup \left(\bigcup A^{(\alpha)} \setminus \{w_2\} \right) \subseteq \bigcup A^{(\nu_2)},$$

so that $\bigcup A^{(\alpha)} = \bigcup A^{(\nu_2)}$, contradicting again the greediness of the process, since $A_{\nu_2} = A = A_\alpha$ and $\nu_2 < \alpha$. $\qquad \square$

3.4   Shadow Processes

Let $\mathcal{H}$ and $\widehat{\mathcal{H}}$ be Π-processes complying with the same Π-board, and let Σ and $\widehat{\Sigma}$ be, respectively, their final partitions. In this section we address the following question:

> *Which properties do the processes* $\mathcal{H}$ *and* $\widehat{\mathcal{H}}$ *have to share in order that* Σ *and* $\widehat{\Sigma}$ *imitate each other?*

Such a question is of course of basic importance to the satisfiability problem for MLSP-like theories, since, by Corollary 2.29, if Σ and $\widehat{\Sigma}$ imitate each other, then they satisfy the same MLSP-formulae.[10]

We shall see that a sufficient condition for Σ and $\widehat{\Sigma}$ to imitate each other is that their respective Π-processes $\mathcal{H}$ and $\widehat{\mathcal{H}}$ are synchronized in the sense detailed by the following definition:

Definition 3.21 (Shadow process). *Let* $\mathcal{H} := \langle (\Pi_\mu)_{\mu \leqslant \xi}, (\bullet), \mathcal{T} \rangle$ *be a* Π*-process with grand-event map* $\mathsf{GE}(\cdot)$ *and edge-activation map* $\mathsf{EA}(\cdot)$. *Likewise, let* $\widehat{\mathcal{H}} := \langle (\widehat{\Pi}_\alpha)_{\alpha \leqslant \widehat{\xi}}, [\bullet], \mathcal{T} \rangle$ *be a* Π*-process with grand-event map* $\widehat{\mathsf{GE}}(\cdot)$ *and edge-activation map* $\widehat{\mathsf{EA}}(\cdot)$.

We say that $\widehat{\mathcal{H}}$ *is a* SHADOW PROCESS *of* $\mathcal{H}$, *if there exists an order-preserving partial map* $\varsigma \colon \{0, 1, \ldots, \widehat{\xi}\} \nrightarrow \{0, 1, \ldots, \xi\}$ *(called* SYNCHRONIZATION MAP, *and also denoted by* $\alpha \mapsto \mu_\alpha$) *such that the following conditions are fulfilled:*

(a) edge activation:

- $\widehat{\mathsf{EA}}[\mathcal{P}(\Pi) \times \Pi] \subseteq \mathsf{dom}(\varsigma)$, *and*
- $\mu_{\widehat{\mathsf{EA}}(B,q)} = \mathsf{EA}(B, q)$, *for every* $B \subseteq \Pi$ *and* $q \in \Pi$;

(b) distribution:

- $\Delta^{[\alpha]}(q) \neq \emptyset \iff \Delta^{(\mu_\alpha)}(q) \neq \emptyset$, *for every* $q \in \Pi$ *and* $\alpha \in \mathsf{dom}(\varsigma)$;

(c) last distribution:

- $\widehat{\mathsf{GE}}[\mathcal{P}(\Pi)] \subseteq \mathsf{dom}(\varsigma)$,
- $\mu_{\widehat{\mathsf{GE}}(B)} = \mathsf{GE}(B)$, *for every* $B \subseteq \Pi$, *and*
- $\bigcup B^{[\bullet]} \in \Delta^{[\widehat{\mathsf{GE}}(B)]}(q) \iff \bigcup B^{(\bullet)} \in \Delta^{(\mathsf{GE}(B))}(q)$, *for every* $B \subseteq \Pi$ *and* $q \in \Pi$. ∎

Condition (a) in Definition 3.21 states that the edge-activation events of $\mathcal{H}$ and $\widehat{\mathcal{H}}$ must be synchronized, in the sense that they must occur with the same order in both Π-processes. Condition (b) asserts that, at each pair of synchronized steps, the targets that get new elements must be exactly the same in both Π-processes. Finally, condition (c) affirms that also all grand-events must be synchronized in both Π-processes, and, additionally, at such steps the maximal set must be assigned to the same target.

The shadow relationship among two Π-processes $\mathcal{H}$ and $\widehat{\mathcal{H}}$ generates also a synchronization among the Π-boards induced by the partitions in $\mathcal{H}$ and $\widehat{\mathcal{H}}$; for details, see Exercise 3.14. Such synchronization will be particularly useful in connection with the pumping mechanism of Section 5.4.

Example 3.22. Consider the transitive partition $\Sigma := \{ q_1^{(\bullet)}, q_2^{(\bullet)}, q_3^{(\bullet)} \}$, with

$$q_1^{(\bullet)} := \{ \emptyset^{2k} \mid k \in \omega \}, \quad q_2^{(\bullet)} := \{ \emptyset^{2k+1} \mid k \in \omega \}, \quad q_3^{(\bullet)} := \{ \{ \emptyset^{2k} \mid k \in \omega \} \}.$$

The syllogistic Π-board $\mathcal{G}$ induced by Σ is

---

[10] The finite enumeration construct $\{\cdot, \ldots, \cdot\}$ and the finiteness predicate $Finite(\cdot)$ of MLSSPF will be dealt with with additional 'external' conditions in Chapters 4 and 5.

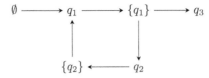

where $\Pi := \{q_1, q_2, q_3\}$.

A formative Π-process for Σ, denoted by $\mathcal{H}$, is

$$
\begin{pmatrix} q_1^{(\mu)} \\ q_2^{(\mu)} \\ q_3^{(\mu)} \end{pmatrix}_{\mu \leqslant \omega+1} = \left(\begin{array}{c|c|c|c|c|c|c|c} \emptyset & \{\emptyset\} & \{\emptyset\} & \{\emptyset, \emptyset^2\} & \{\emptyset, \emptyset^2\} & \cdots & \{\emptyset, \emptyset^2, \emptyset^4, \ldots\} & \{\emptyset, \emptyset^2, \emptyset^4, \ldots\} \\ \emptyset & \emptyset & \{\emptyset^1\} & \{\emptyset^1\} & \{\emptyset^1, \emptyset^3\} & \cdots & \{\emptyset^1, \emptyset^3, \emptyset^5, \ldots\} & \{\emptyset^1, \emptyset^3, \emptyset^5, \ldots\} \\ \emptyset & \emptyset & \emptyset & \emptyset & \emptyset & \cdots & \emptyset & \{\{\emptyset, \emptyset^2, \emptyset^4, \ldots\}\} \end{array} \right),
$$

with trace and history

$$
\begin{pmatrix} A_\mu \\ T_\mu \end{pmatrix}_{\mu < \omega+1} = \left(\begin{array}{c|c|c|c|c|c} \emptyset & \{q_1\} & \{q_2\} & \{q_1\} & \cdots & \{q_1\} \\ \{q_1\} & \{q_2\} & \{q_1\} & \{q_2\} & \cdots & \{q_3\} \end{array} \right).
$$

The Π-process $\mathcal{H}$ consists of the execution of the initial step $\emptyset \to q_1$ (at which the element $\emptyset$ is assigned to q_1), followed by ω iterations of the cycle

$$\{q_1\} \to q_2 \to \{q_2\} \to q_1 \to \{q_1\} \tag{3.18}$$

(which assign to q_1 all the elements $\emptyset^{2k+2}$ and to q_2 all the elements $\emptyset^{2k+1}$, for $k \geqslant 0$). Finally, an execution of step $\{q_1\} \to q_3$ assigns, as an element, the set $\{\emptyset, \emptyset^2, \emptyset^4, \ldots\}$ to q_3.

Notice also that $\emptyset^{(\bullet)} \in q_1^{(\bullet)}$ and $\{q_1\}^{(\bullet)} \in q_3^{(\bullet)}$. Thus, we say that the places q_1 and q_3 are the *principal targets* of the nodes $\emptyset$ and $\{q_1\}$, respectively.

We describe the above situation by saying that the Π-process $\mathcal{H}$ is a *realization of the Π-board $\mathcal{G}$ extended with the principal target information* $\{\langle \emptyset, q_1\rangle, \langle \{q_1\}, q_3\rangle\}$.

Are the ω iterations of the cycle (3.18) necessary in connection with MLS-satisfiability, or is there some simpler realization of the same extended Π-board G?

The answer to the latter question is that the ω iterations of the cycle (3.18) are not necessary. Indeed, consider the following Π-process

$$
\begin{pmatrix} q_1^{[\alpha]} \\ q_2^{[\alpha]} \\ q_3^{[\alpha]} \end{pmatrix}_{\alpha \leqslant 4} = \left(\begin{array}{c|c|c|c|c} \emptyset & \{\emptyset\} & \{\emptyset\} & \{\emptyset, \emptyset^2\} & \{\emptyset, \emptyset^2\} \\ \emptyset & \emptyset & \{\emptyset^1\} & \{\emptyset^1\} & \{\emptyset^1\} \\ \emptyset & \emptyset & \emptyset & \emptyset & \{\{\emptyset, \emptyset^2\}\} \end{array} \right),
$$

with trace and history

$$
\begin{pmatrix} \widehat{A}_\beta \\ \widehat{T}_\beta \end{pmatrix}_{\beta < 4} = \left(\begin{array}{c|c|c|c} \emptyset & \{q_1\} & \{q_2\} & \{q_1\} \\ \{q_1\} & \{q_2\} & \{q_1\} & \{q_3\} \end{array} \right).
$$

It is an easy matter to check that the final partition $\widehat{\Sigma} := \{q_1^{[\bullet]}, q_2^{[\bullet]}, q_3^{[\bullet]}\}$, where

$$q_1^{[\bullet]} := \{\emptyset, \emptyset^2\}, \quad q_2^{[\bullet]} := \{\emptyset^1\}, \quad q_3^{[\bullet]} := \{\{\emptyset, \emptyset^2\}\},$$

is a realization of the Π-board $\mathcal{G}$ extended with the principal target information $\{\langle \emptyset, q_1 \rangle, \langle \{q_1\}, q_3 \rangle\}$. Therefore, the two partitions Σ and $\widehat{\Sigma}$ satisfy the same MLS-formulae. In fact, $\widehat{\Sigma}$ is a shadow process of Σ, with synchronization map

$$\varsigma := \{\langle 0, 0 \rangle, \langle 1, 1 \rangle, \langle 2, 2 \rangle, \langle 3, \omega \rangle, \langle 4, \omega + 1 \rangle\}.$$

Figure 3.4 shows the sequences of Π-boards induced by the intermediate partitions, after each move, of the Π-processes $\mathcal{H}$ (on the left) and $\widehat{\mathcal{H}}$ (on the right). Newly activated edges are bold-faced. Synchronization of the two sequences is manifest (see Exercise 3.14). ∎

As anticipated, given two Π-processes, where one of which is the shadow of the other, the final partition of the shadow process imitates the final partition of the other process. This is the content of the following important theorem.

Theorem 3.23 (Shadow Theorem). *Let* $\mathcal{H} := \left\langle (\Pi_\mu)_{\mu \leqslant \xi}, (\bullet), \mathcal{T} \right\rangle$ *and* $\widehat{\mathcal{H}} := \left\langle (\widehat{\Pi}_\alpha)_{\alpha \leqslant \widehat{\xi}}, [\bullet], \mathcal{T} \right\rangle$ *be* Π-*processes such that* $\widehat{\mathcal{H}}$ *is a shadow process of* $\mathcal{H}$*. Then,* $\Pi^{[\bullet]}$ *imitates* $\Pi^{(\bullet)}$.

Proof. We show that $\Pi^{[\bullet]} = \widehat{\Pi}_{\widehat{\xi}}$ imitates $\Pi^{(\bullet)} = \Pi_\xi$ via the bijection $q^{(\bullet)} \overset{\beta}{\longmapsto} q^{[\bullet]}$. In view of Remark 2.27, it is enough to prove that the conditions

(i) $\bigcup B^{[\bullet]} \in q^{[\bullet]}$ if and only if $\bigcup B^{(\bullet)} \in q^{(\bullet)}$, and

(ii) if $\mathcal{P}^*\left(B^{[\bullet]}\right) \cap q^{[\bullet]} \neq \emptyset$, then $\mathcal{P}^*\left(B^{(\bullet)}\right) \cap q^{(\bullet)} \neq \emptyset$

hold for every $B \subseteq \Pi$ and $q \in \Pi$.[11]

Concerning (i), we have:

$$\begin{aligned}
\bigcup B^{[\bullet]} \in q^{[\bullet]} &\Longleftrightarrow \bigcup B^{[\bullet]} \in q^{[\widehat{GE}(B)+1]} \setminus q^{[\widehat{GE}(B)]} & \text{(by the definition of the map } \widehat{GE}) \\
&\Longleftrightarrow \bigcup B^{[\bullet]} \in \Delta^{[\widehat{GE}(B)]}(q) \\
&\Longleftrightarrow \bigcup B^{(\bullet)} \in \Delta^{(GE(B))}(q) & \text{(by the last-distribution condition)} \\
&\Longleftrightarrow \bigcup B^{(\bullet)} \in q^{(GE(B)+1)} \setminus q^{(GE(B))} \\
&\Longleftrightarrow \bigcup B^{(\bullet)} \in q^{(\bullet)} & \text{(by the definition of the map } GE).
\end{aligned}$$

Regarding (ii), let us assume that $\mathcal{P}^*\left(B^{[\bullet]}\right) \cap q^{[\bullet]} \neq \emptyset$, for a node $B \subseteq \Pi$ and place $q \in \Pi$. Then, putting $\widehat{\nu} := \widehat{EA}(B, q)$, we have $\widehat{\nu} < \widehat{\xi}$. Therefore, since $\mu_{\widehat{\xi}} = \xi$ (as shown in Exercise 3.13), we have $\mu_{\widehat{\nu}} = EA(B, q) < \xi$, so that $\mathcal{P}^*\left(B^{(\bullet)}\right) \cap q^{(\bullet)} \neq \emptyset$.

We conclude that $\Pi^{[\bullet]}$ imitates $\Pi^{(\bullet)}$, as claimed. □

---

[11]In fact, as can be easily seen, the implication in (ii) can be reversed, and therefore one can prove the stronger result

$$\mathcal{P}^*\left(B^{[\bullet]}\right) \cap q^{[\bullet]} \neq \emptyset \qquad \Longleftrightarrow \qquad \mathcal{P}^*\left(B^{(\bullet)}\right) \cap q^{(\bullet)} \neq \emptyset,$$

for every $B \subseteq \Pi$ and $q \in \Pi$.

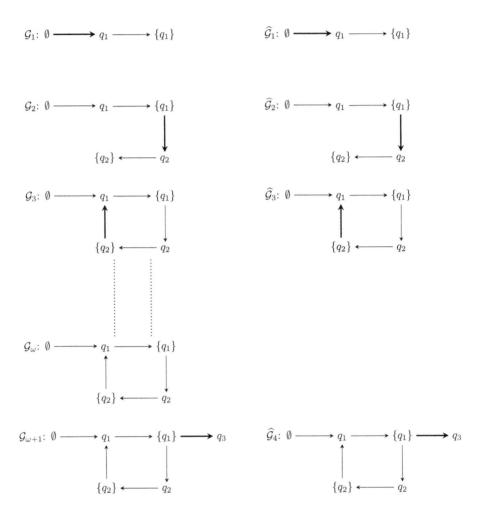

Figure 3.7: Synchronicity of the Π-boards induced by the Π-processes $\mathcal{H}$ (on the left) and $\widehat{\mathcal{H}}$ (on the right).

EXERCISES

Exercise 3.1. *Prove that the syllogistic board induced by a finite partition with an infinite domain must contain some cycle.*

Exercise 3.2. *Let Σ_∞ be the partition defined by (3.4). Prove the following statements:*

(a) Σ_∞ *is transitive;*

(b) Σ_∞ *imitates the partition Σ_0 defined by (3.1);*

(c) Σ_∞ *has a formative process that is a shadow of the formative process (3.2) and whose trace is the sequence (3.5).*

Exercise 3.3. *Let Σ be a $\mathcal{P}$-partition, with special block $\bar{\sigma}$, complying with a Π-board $\mathcal{G}$ via a map $q \mapsto q^{(\bullet)}$. Also, let $\bar{q} \in \Pi$ be such that $\bar{q}^{(\bullet)} = \bar{\sigma}$.*
   Show that $\mathcal{T}(B) = \emptyset$, for each node $B \subseteq \Pi$ such that $\bar{q} \in B$, where $\mathcal{T}$ is the target function of $\mathcal{G}$.

Exercise 3.4. *Let $\left(\Pi_\mu\right)_{\mu \leqslant \xi}$ be a formative process. Prove that, for all ordinals $\nu < \mu \leqslant \xi$ and each block $\sigma \in \Pi_\xi$, the following properties hold:*

(i) $\sigma^{(\mu)} = \sigma \cap \bigcup \Pi_\mu$;

(ii) $\Pi_\mu = \left\{ \tau^{(\mu)} \mid \tau \in \Pi_\xi \right\}$;

(iii) $\bigcup \Pi_\nu \subsetneq \bigcup \Pi_\mu$;

(iv) $\bigcup \Pi_\mu$ *frames* $\bigcup \Pi_\nu$;

(v) $\bigcup \Pi_\lambda = \bigcup \{ \bigcup \Pi_\gamma \mid \gamma < \lambda \}$, *for every limit ordinal $\lambda \leqslant \xi$.*

Exercise 3.5. *Let $\left(\Pi_\mu\right)_{\mu \leqslant \xi}$ be a formative process and let $e \in \Pi_\xi$. Show that the least ordinal ν such that $e \in \bigcup \Pi_\nu$ is a successor ordinal.*

Exercise 3.6. *Prove Lemma 3.9.*

Exercise 3.7. *Prove the set inclusions (3.14), (3.15), and (3.16) in the proof of the Trace Theorem.*

Exercise 3.8. *Let $\left\langle \left(\Pi_\mu\right)_{\mu \leqslant \xi}, (\bullet), \mathcal{T}, \mathcal{R} \right\rangle$ be a colored Π-process with edge-activation map $\mathsf{EA} \colon \mathcal{P}(\Pi) \times \Pi \to \{0, 1, \ldots, \xi\}$. Show that, for every $B \subseteq \Pi$ and $q \in \Pi$, we have:*

$$\mathsf{EA}(q) = \min\{\mathsf{EA}(B, q) \mid B \subseteq \Pi\}.$$

Exercise 3.9. *Prove Remark 3.16.*

Exercise 3.10. *Prove Lemma 3.19.*

Exercise 3.11. *For any fixed colored Π-board $\langle \mathcal{T}, \mathcal{R} \rangle$, prove that the shadow relationship among colored Π-processes is an equivalence relation.*

Exercise 3.12. *Show that in a colored Π-process with edge-activation map $\mathsf{EA}(\cdot)$ the following set inclusion holds:*

$$\mathsf{EA}[\Pi] \subseteq \mathsf{EA}[\mathcal{P}(\Pi) \times \Pi]\,.$$

Exercise 3.13. *Let $\mathcal{H}$ and $\widehat{\mathcal{H}}$ be colored Π-processes of length ξ and $\widehat{\xi}$, respectively. Show that if $\widehat{\mathcal{H}}$ is a shadow process of $\mathcal{H}$, with synchronization map ς, then we have:*

$$\widehat{\xi} \in \mathsf{dom}(\varsigma), \qquad \xi \in \mathsf{ran}(\varsigma), \qquad \mu_{\widehat{\xi}} = \xi\,.$$

Exercise 3.14. *Let $\mathcal{H} := \left\langle (\Pi_\mu)_{\mu \leqslant \xi}, (\bullet), \mathcal{T} \right\rangle$ and $\widehat{\mathcal{H}} := \left\langle (\widehat{\Pi}_\alpha)_{\alpha \leqslant \widehat{\xi}}, [\bullet], \mathcal{T} \right\rangle$ be Π-processes such that $\widehat{\mathcal{H}}$ is a shadow process of $\mathcal{H}$ via the synchronization map $\varsigma \colon \left\{0, 1, \ldots, \widehat{\xi}\right\} \twoheadrightarrow \left\{0, 1, \ldots, \xi\right\}$.*

  (i) *Show that $0, \widehat{\xi} \in \widehat{\mathsf{EA}}[\mathcal{P}(\Pi) \times \Pi]$ and $0, \xi \in \mathsf{EA}[\mathcal{P}(\Pi) \times \Pi]$.*

  (ii) *Let, additionally, $\widehat{\mathsf{EA}}[\mathcal{P}(\Pi) \times \Pi] = \{\alpha_0, \ldots, \alpha_k\}$ and $\mathsf{EA}[\mathcal{P}(\Pi) \times \Pi] = \{\mu_{\alpha_0}, \ldots, \mu_{\alpha_k}\}$, with*

$$0 = \alpha_0 < \ldots < \alpha_k = \widehat{\xi} \qquad and \qquad 0 = \mu_{\alpha_0} < \ldots < \mu_{\alpha_k} = \xi\,.$$

  Prove that, for all i, α, and μ, where $i \in \{0, 1, \ldots, k-1\}$, $\alpha_i < \alpha \leqslant \alpha_{i+1}$, and $\mu_{\alpha_i} < \mu \leqslant \mu_{\alpha_{i+1}}$, the partitions $\widehat{\Pi}_\alpha$ and Π_μ induce isomorphic Π-boards.

Exercise 3.15. *Show that there is a unique greedy formative process for the transitive partition $\Sigma := \left\{q_1^{(\bullet)}, q_2^{(\bullet)}, q_3^{(\bullet)}\right\}$ of Example 3.22, where*

$$q_1^{(\bullet)} := \left\{\emptyset^{2k} \mid k \in \omega\right\}, \quad q_2^{(\bullet)} := \left\{\emptyset^{2k+1} \mid k \in \omega\right\}, \quad q_3^{(\bullet)} := \left\{\left\{\emptyset^{2k} \mid k \in \omega\right\}\right\}\,.$$

Exercise 3.16. *Let $\mathcal{P}_\omega(s)$ denote the collection of the finite subsets of s, i.e.,*

$$\mathcal{P}_\omega(S) := \left\{s \in \mathcal{P}(S) \mid |s| < \aleph_0\right\},$$

and let Φ be the formula

$$\mathcal{P}_\omega(x) = x.$$

Is Φ satisfiable? In the event that Φ were satisfiable, provide a model M for it and a formative process for its Venn $\mathcal{P}$-partition.

Part II

Applications

Chapter 4

Decidability of MLSSP

Towards a proof of the decidability of **MLSSP**, there are two fundamental goals to achieve. The first one consists in finding a shadow process that is good enough to create an assignment that L-simulates the original one and, therefore, using Lemma 2.24, also good enough to create a model for the original formula.

In order to make effective a non-deterministic search of a model, we need to bound the search space. More specifically, this bound must be related to the formula in such a way as to assert that whenever there is a model, there must be one that can be constructed through a process whose length does not exceed a fixed number, which depends only on the formula. In this way, when we attempt to decide whether a formula is or is not satisfiable, it is sufficient to perform a non-deterministic check of all models whose rank does not exceed a fixed number or, in other terms, a check of all models that can be constructed with formative processes not exceeding a fixed length.

The above discussion explains why the second goal must consist in bounding the length of this shadow process (and at the same time the rank of the assignment, as we shall see) by a number that strictly depends on the size of the formula, more specifically, on the number of distinct variables occurring in the formula. The achievement of such claims will yield, as a by-product, a procedure that decides, for a given formula, whether it is satisfiable.

Throughout the present chapter, all partitions resulting from a formative process are to be considered $\mathcal{P}$-partitions.

4.1 Highlights of a Decision Test for MLSSP

The decidability of **MLSSP** has been established in [Can91, COU02] by proving that **MLSSP** enjoys a small model property. This, we recall, consists in showing that there exists a computable function $c : \omega \to \omega$ such that any satisfiable **MLSSP**-formula Φ with m distinct variables has a model of rank bounded by $c(m)$. Therefore, to test the satisfiability of an **MLSSP**-formula Φ involving the variables $\text{Vars}(\Phi)$, it is sufficient to check all possible assignments of rank less than $c(|\text{Vars}(\Phi)|)$. The following high level algorithm uses this kind of approach:

> **procedure** SatifisfiabilityTestMLSSP(Φ);
>> **for each** set assignment M over V_Φ, whose rank is bounded by $c(|V_\Phi|)$ **do**
>>> **if** M satisfies Φ **then**
>>>> **return** "Φ is satisfiable";

```
        end if;
      end for;
      return "Φ is unsatisfiable";
  end procedure;
```

The effectiveness of the above procedure follows immediately by observing that

- there are only finitely many set assignments over $\mathrm{Vars}(\Phi)$ of rank not exceeding $c(|\mathrm{Vars}(\Phi)|)$ (and these can be effectively generated), and

- it can be effectively verified whether any such set assignment satisfies or not the formula Φ.

4.2 Shadow Processes and the Decidability of MLSSP

In this section, we start from a given MLSSP-formula Φ, a model for Φ, and a related partition created by a greedy process. Arguing from Definition 1.22, we can assume without loss of generality that such a partition is actually a $\mathcal{P}$-partition. Our goal is to create in detail, as we anticipated in Section 3.4, a new process which is a shadow process of the given one.

Specifically, we decide which moves, inside a formative process, are essential in order to satisfy MLSSP-literals, but more than this, we shall show that it is possible to construct a process, which will be a shadow process, using just the selected steps, which is probably the most intriguing aspect of this section. Indeed, when we select the relevant steps of a process, we are still far from being in the condition to create a process, as we shall explain in the following.

In other terms we have to decide which steps (called "Salient Steps") cannot be removed without affecting the capability of the resulting transitive partition to generate a model for the given formula. In order to prune the original process, a procedure has been created to select the salient steps and to create a new formative process which generates a new transitive partition.

It has the same structural properties of the original one, where the structural properties are set having in mind which kinds of problems we are going to solve. In our case, we deal with MLSSP-formulae. This means that we need to find a new $\mathcal{P}$-partition which L-simulates the original one in as few steps as possible. Indeed, for each finite transitive partition Σ and integer constant $L \geqslant 0$, we shall prove that there exists a transitive partition of finite bounded rank which L-simulates Σ. This result has been established in [COU02] by means of a procedure called imitate that we shall see in detail in Section 4.2.4. We proceed through the following steps:

- we select the salient steps and give a picture of a shadow process (Section 4.2.1);

- we find a natural number ρ_0 which will play a fundamental role in order to bound the length of the desired shadow process (Section 4.2.3);

- using salient steps, we show a procedure which creates a new process (Section 4.2.4);

- we prove that such a process is a shadow process of the original one (this fact, by the Shadow Theorem, implies that the partition resulting from the shadow process simulates the original one) (Section 4.2.6);

- we prove that the shadow process has length not exceeding a natural number that depends only on the number of variables of the formula (Section 4.2.7);

- using all the above results, we conclude that MLSSP is decidable (Section 4.2.8).

4.2.1 Salient Steps and Shadow Process

We start with a description of the strategy for the selection of salient steps of a greedy formative Π-process $\langle (\Pi_\mu)_{\mu \leq \xi}, (\bullet), \mathcal{T} \rangle$, whose sequence of moves is $(A_\mu)_{\mu < \xi}$.

A trivial approach could be simply to select all the steps until the formative process produces a partition which L-simulates the original one, which would ensure that the model satisfies the same MLSSP models as the original does. Unfortunately, this procedure is not bounded by a number which depends only on the formula. This means that we cannot, in general, use all the steps of the original process, but just a reduced number, for example those steps that create the structure of the Π-boards. Now, the other claim of this section enters the scene. That is, how to make this selection effective. In other terms, how to guarantee, when we abandon the original process, that we can still create a process with the selected moves. Here the role of the number ρ becomes crucial.

The main scope of the shadow process is to simulate as far as possible the original process, at least, by creating edges between places and $\mathcal{P}$-nodes when the original does.

Suppose that, at step μ_i, the original process creates four edges, or, in other terms, it distributes new elements to four targets so far never used. We expect that, at step i, the shadow process does the same. But the shadow process is not the same process so, in general, it could happen that this operation cannot be done, for example because $\mathcal{P}^*(\widehat{B}_i)$ has no more than three new elements. How can such a pathology be avoided? What we really need is that, at each step, the shadow process, even if it abandons the original process, is at least capable of creating the same edges as the original process does. This means that it must have at least as many new elements as the number of the total amount of blocks.

Suppose this number is n. Unfortunately, in the subsequent step we could fall into the same pathology. So we ask the shadow process to have enough new elements at each step in order to create at least the same amount of edges as the original process does. In a sense, we ask that the capability to distribute n elements propagates, as a "chain reaction". What we know is that whenever a new element is distributed to a block, if it possesses h elements, every $\mathcal{P}$-node to which it belongs has at least 2^{h-1} new elements to distribute. So we need a number which is able to produce at least n copies of itself, in other terms a number ρ such that $2^{\rho-1} > n\rho$, which is exactly our requested feature for ρ. Indeed, as soon as a block has reached this size, all the $\mathcal{P}$-nodes to which it belongs have more than $n\rho$ new elements, so no matter how many edges they have to create in the following step, we are sure that they can, since new elements are more than the total amount of the blocks. Although they must do even more, being as they have to put all those blocks in the condition to do the same, since they must distribute ρ new elements to each target. Even if the targets to activate would be as many as possible, that means n, this can be done since new elements have not to be just n but $n\rho$. This, roughly speaking, is the role played by the number ρ.

We must characterize what components of the input process should be regarded as being *salient*. Following [COU02, Definition 7.1], we do as follows:

Definition 4.1. *The ν-th step (with $0 \leqslant \nu < \xi$) of a greedy formative process $(\Pi_\mu)_{\mu \leqslant \xi}$ is said to be* SALIENT *if it meets one of the following conditions:*

1. *each block in $A_\nu^{(\nu)}$ is small. The threshold size below which a block is said to be* SMALL *is a number ϱ such that $2^{\varrho-1} > \varrho \cdot |\Pi|$ (and $\varrho > L$, where L is fixed).*

2. *one of the blocks $t^{(\nu)}$, with $t \in \mathcal{T}(A_\nu)$, that are about to receive new elements, either*

   (a) *is still small; or*

   (b) *never got anything from the delivering partition, i.e., $\mathcal{P}^*\left(A_\nu^{(\nu)}\right) \cap t^{(\nu)} = \emptyset$ (edge activation);*

3. $\mathsf{GE}(A_\nu) = \nu$.   ■

Remark 4.2. Notice that:

- the above criteria for saliency are not mutually exclusive, e.g., $A_\nu^{(\nu)}$ could be *small* and, in the meanwhile, one of the blocks $t^{(\nu)}$, with $t \in \mathcal{T}(A_\nu)$ is still small. In this case the step could be selected for two different criteria. Nevertheless,

- to justify inclusion of the ν-th step among the salient ones, only one criterion will be considered, according to the priority scale $1, 3, 2$.   ■

Then the SALIENT SEQUENCE (WITH RESPECT TO THE THRESHOLD ϱ) of $\left\langle (\Pi_\mu)_{\mu \leqslant \xi}, (\bullet), \mathcal{T} \right\rangle$ is the sequence $\{\mu_i\}_{i \leqslant \ell}$ such that

- $\mu_0, \mu_1, \ldots, \mu_{\ell-1}$ are the salient steps, and

- $\mu_0 < \mu_1 < \ldots < \mu_\ell = \xi$

(hence, in particular, $\mu_0 = 0$).

For each i, the number in $\{1, 2, 3\}$ is used in order to include the μ_i-th step among salient ones. This is called the *justification* for μ_i. The intended meaning of the numbers $\{1, 2, 3\}$ is the reason for which the step μ_i has been selected as salient step.

Depending on whether its justification is 1, 2, or 3, each μ_i can be classified as being *scant,*\textbf *innovative*, or *finishing*—cf. Figure 4.1: "innovative moves" μ_i are called so to emphasize the fact that something changes in the Π-board, which means that at least one edge has been added or that at least one block has turned out to be no longer small. Figure 4.2 describes the main features of the procedure imitate, whose details will be displayed in the next section: it is understood that the i-th step of the mimicking process makes use of the same A_{μ_i} which is exploited at the μ_i-th step by the original process, and hence its delivering partition is $\widehat{A}_{\mu_i}^{[i]}$.

4.2.2   Procedure Imitate: The Algorithm

The procedure imitate constructs a shadow process $\left(\widehat{\Pi}_i\right)_{i \leqslant \ell}$, by suitably mimicking the sequence of 'salient' actions $(\langle \Pi_{\mu_j}, \Pi_{\mu_j+1} \rangle)_{j < \ell}$ of the process $(\Pi_\mu)_{\mu \leqslant \xi}$, starting from an empty partition $\widehat{\Pi}_0$. This results in a sequence of actions $\left(\langle \widehat{\Pi}_j, \widehat{\Pi}_{j+1} \rangle\right)_{j < \ell}$, based on the sequence of moves $\left(\widehat{A}_j\right)_{j < \ell}$, where $\widehat{A}_j = A_{\mu_j}$, for $0 \leqslant j < \ell$.

Let Π be any set of places such that $|\Pi| = |\Pi_\xi|$ (and $\Pi \cap \mathcal{P}(\Pi) = \emptyset$). Also, let $q \mapsto q^{(\bullet)}$ be a bijection from Π onto Π_ξ and let $\mathcal{T} \in \mathcal{P}(\Pi)^{\mathcal{P}(\Pi)}$ be the target function such that

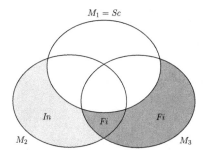

$Sc \, (= M_1)$: scant steps,
$In \, (= M_2 \backslash M_1 \backslash M_3)$: innovative steps,
$Fi \, (= M_3 \backslash M_1)$: finishing steps.

Accordingly, the set of all salient steps is
$$Sal = M_1 \cup M_2 \cup M_3 = Sc \cup In \cup Fi,$$
where
$$Sc \cap In = Sc \cap Fi = In \cap Fi = \emptyset.$$

Figure 4.1:   Classification of salient steps.

$\mathcal{T}(B) = \{q \in \Pi \mid q^{(\bullet)} \cap \mathcal{P}^*(B^{(\bullet)}) \neq \emptyset\}$, for every $B \subseteq \Pi$, so that the transitive partition Π_ξ complies, via $q \mapsto q^{(\bullet)}$, with the Π-board $\mathcal{G}$, whose target function is $\mathcal{T}$. Likewise, let $\widehat{\Pi}$ be any set of places such that $|\widehat{\Pi}| = |\Pi|$ (and $\widehat{\Pi} \cap \mathcal{P}(\widehat{\Pi}) = \emptyset$) and let $q \mapsto \widehat{q}$ be a bijection from Π onto $\widehat{\Pi}$.[1]

Then, there exists a bijection $\widehat{q} \mapsto \widehat{q}^{[\bullet]}$ from $\widehat{\Pi}$ onto $\widehat{\Pi}_\ell$ such that $\widehat{\Pi}_\ell$ complies, via $\widehat{q} \mapsto \widehat{q}^{[\bullet]}$, with the (shadow) $\widehat{\Pi}$-board $\widehat{\mathcal{G}}$ with target function the map $\widehat{\mathcal{T}} \in \mathcal{P}(\widehat{\Pi})^{\mathcal{P}(\widehat{\Pi})}$ defined by $\widehat{\mathcal{T}}(B) := \{\widehat{q} \in \widehat{\Pi} \mid \widehat{q}^{[\bullet]} \cap \mathcal{P}^*(\widehat{B}^{[\bullet]}) \neq \emptyset\}$, for $\widehat{B} \subseteq \widehat{\Pi}$.

In analogy to (3.13), for all $0 \leqslant i \leqslant \ell$ and $\widehat{\tau} \in \widehat{\Pi}_\ell$, we put

$$\widehat{\tau}^{[i]} \quad := \quad \text{the unique } \widehat{\sigma} \in \widehat{\Pi}_i \text{ such that } \widehat{\sigma} \subseteq \widehat{\tau}, \text{ if any exists, else } \emptyset, \tag{4.1}$$

and, for all $0 \leqslant i \leqslant \ell$, $0 \leqslant j < \ell$, $\widehat{p} \in \widehat{\Pi}$, $\widehat{B} \subseteq \widehat{\Pi}$, we designate by $\widehat{p}^{[i]}, \widehat{B}^{[i]}, \Delta^{[j]}(\widehat{p}), \widehat{T}^{[j]}$, respectively, the unique sets such that

$$
\begin{aligned}
\widehat{p}^{[i]} &= (\widehat{p}^{[\bullet]})^{[i]} \\
\widehat{B}^{[i]} &= \{\widehat{q}^{[i]} : \widehat{q} \in \widehat{B}\} \\
\Delta^{[j]}(\widehat{p}) &= \widehat{p}^{[j+1]} \backslash \bigcup \widehat{\Pi}^{[j]} \\
\widehat{T}_j &= \{\widehat{p} \in \widehat{\Pi} \mid \Delta^{[j]}(\widehat{p}) \neq \emptyset\},
\end{aligned}
$$

where $\widehat{q} \mapsto \widehat{q}^{[\bullet]}$ is the bijection depicted above. Then, more in general, we have

4.2.3   Estimation of ϱ

First we state a technical lemma which will be useful in the prosecution.

Lemma 4.3. *Let $(\Pi_\mu)_{\mu \leqslant \xi}$ be a weak formative process such that $\Pi_\xi \not\subseteq \{\{\emptyset\}\}$ and whose trace is $(A_\nu)_{\nu < \xi}$. Then the following condition holds, for $q \in \Pi$, $B \subseteq \Pi$, $\mu \leqslant \xi$, $\nu \leqslant \mu$, and $\nu < \xi$:*

---

[1] Observe that there are no mathematical reasons to assign different names to Π and $\widehat{\Pi}$ (and to the objects related to them, such as $\widehat{q}, \widehat{B}, \mathcal{D}$, etc.), since, essentially, they are the same set. These can be considered as a collection of empty boxes, so, at the very beginning of the process, they are indistinguishable. The reason for different names is in what they are going to be along the process. Indeed, our motivation for using two different notations for the same object is to stress the fact that Π and $\widehat{\Pi}$ are the starting points of (possibly) different formative processes, namely, the original one and its shadow counterpart.

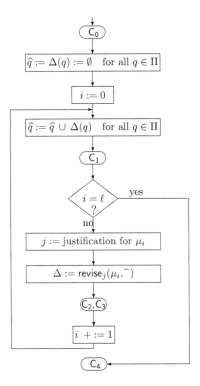

Figure 4.2: Flow-chart of the procedure shortenGame.

- $\emptyset \notin A_\nu^{(\mu)}$ *(hence $A_\nu^{(\mu)}$ is a partition),*

- $q^{(\nu)} \subseteq q^{(\mu)}$, *and $B^{(\mu)}$ frames $B^{(\nu)}$*
  (more accurately stated, $B^{(\mu)} \setminus \{\emptyset\}$ frames $B^{(\nu)} \setminus \{\emptyset\}$),
  and hence $\bigcup B^{(\nu)} \subseteq \bigcup B^{(\mu)}$,

- rk $q^{(\mu)} \leqslant \mu$,

- rk $B^{(\mu)} \leqslant \mu + 1$;

Proof. Straightforward.   □

By Lemma 2.24, the capability of L-simulating a given finite transitive partition Σ by means of another partition $\widehat{\Sigma}$, created through a shadow process and whose rank is bounded by a computable function in L and $|\Sigma|$, is crucial in order to solve the decision problem for collections of literals of the form (4.6).

In Section 4.2.4 we shall describe a non-deterministic procedure, imitate, which carries out this task. More specifically, given a greedy formative process $(\Pi_\mu)_{\mu \leqslant \xi}$ for a transitive partition $\Pi^{(\bullet)}$ (in our minds the Venn regions of a potential model) and given a constant

$\varrho > L$ such that $2^{\varrho-1} > \varrho \cdot |\Pi|$, the procedure imitate will compute a (weak) formative process $(\widehat{\Pi}^{[i]})_{i \leqslant \ell_\varrho}$ for a transitive partition $\widehat{\Pi}^{[\ell_\varrho]}$, which L-simulates $\Pi^{(\bullet)}$. By Lemma 4.2.7 ℓ_ϱ satisfies $\ell_\varrho < \varrho \cdot |\Pi| \cdot 2^{|\Pi|} + 3$.

Let us put $\varrho_0 = \max(\lceil \frac{25}{24} \cdot \lceil \log |\Pi| \rceil + 5 \rceil, L + 1)$. Then we have $\varrho_0 > L$.

The following lemma shows exactly that the above constant satisfies the requested property, which will turn out to be essential in order to make possible the construction of a shadow process.

Lemma 4.4. *Let $y \geqslant 1$, and let $\varrho_y = \lceil \frac{25}{24} \lceil \log y \rceil + 5 \rceil$. Then $2^{\varrho_y - 1} > \varrho_y y$.*

Proof. By elementary calculus, it is immediate to check that

$$2^{\frac{x}{24}+4} > \tfrac{25}{24}x + 6 \,, \text{ for } x \in \mathbb{R} \,.$$

Therefore

$$2^{\lceil \frac{x}{24}+5 \rceil - 1} \geqslant 2^{\frac{x}{24}+4} > \tfrac{25}{24}x + 6 > \lceil \tfrac{25}{24}x + 5 \rceil \,, \text{ for } x \in \mathbb{R} \,.$$

Let $y \geqslant 1$, and put $x = \lceil \log y \rceil$, $\varrho_y = \lceil \frac{25}{24} \lceil \log y \rceil + 5 \rceil$. Then we have

$$2^{\varrho_y - 1} = 2^{\lceil \frac{25}{24}x + 5 \rceil - 1} = 2^{\lceil \frac{x}{24}+5 \rceil - 1} 2^x > \lceil \tfrac{25}{24}x + 5 \rceil y = \varrho_y y \,,$$

which proves the lemma. $\qquad\qquad\qquad\qquad\qquad\qquad\qquad\qquad\qquad\qquad\qquad\qquad \square$

We are now in the condition to assert that $2^{\varrho_0 - 1} > \varrho_0 \cdot |\Pi|$. Hence, what we really promise is that imitate generates formative processes whose length does not exceed

$$\ell_{\varrho_0} < \max \left(\left\lceil \frac{25}{24} \cdot \lceil \log |\Pi| \rceil + 5 \right\rceil, L + 1 \right) \cdot |\Pi| \cdot 2^{|\Pi|} + 3,$$

which we shall show in Section 4.2.7.

Thanks to Lemma 4.3, Lemma 2.24, and Claim **C4** of imitate, we shall have that whenever a formula with $|\Pi|$ syntactical Venn regions has a model, it has a model of rank less than $\max \left(\lceil \frac{25}{24} \cdot \lceil \log |\Pi| \rceil + 5 \rceil, L + 1 \right) \cdot |\Pi| \cdot 2^{|\Pi|} + 3$. In other terms this means that if we want to decide whether a formula is satisfiable or not, it is enough to run the procedure SatisfiabilityTestMLSSP throughout all the assignments created by the formative processes of length l_ϱ generated by the procedure imitate. In detail, the execution of imitate refers, as an oracle, to a greedy formative process of $\Pi^{(\bullet)}$. Should this process not be available, an execution could nevertheless be performed, albeit non-deterministically, to take into consideration all possible response sequences from the oracle; then, at the tip of each branch of the non-deterministic execution tree, one could directly establish whether or not the sequence $(\widehat{\Pi}^{[i]})_{i \leqslant \ell}$ constructed by the procedure imitate L-simulates $\Pi^{(\bullet)}$. Observe that, by Lemma 4.3, we can obtain the same result if we run the procedure SatisfiabilityTestMLSSP throughout all the assignments of rank less than $l_{\rho_0} + 1$.

In view of Lemma 3.9, the above discussion can be summarized as follows.

Lemma 4.5. *Let $(\Pi^{(\mu)})_{\mu \leqslant \xi}$ be a greedy formative process for a transitive partition $\Pi^{(\bullet)}$ and let $L \geqslant 0$. Then there exists a greedy formative process $(\widehat{\Pi}^{[i]})_{i \leqslant \ell}$ for a transitive partition $\widehat{\Pi}$ which L-simulates $\Pi^{(\bullet)}$ and such that*

$$\ell < \max \left(\lceil \tfrac{25}{24} \cdot \lceil \log |\Pi| \rceil + 5 \rceil, L + 1 \right) \cdot |\Pi| \cdot 2^{|\Pi|} + 3.$$

Therefore, by Lemma 4.3, $\mathrm{rk}\ \widehat{\Pi} \leqslant \max\left(\left\lceil \frac{25}{24} \cdot \lceil \log|\Pi| \rceil + 5 \right\rceil, L + 1\right) \cdot |\Pi| \cdot 2^{|\Pi|} + 3.$

The proof of the preceding lemma is postponed to the end of Section 4.2.7.

4.2.4   Procedure Imitate: The Code

For technical reasons, we shall assume that $\Pi^{(\bullet)} \not\subseteq \{\{\emptyset\}\}$. It easily turns out that this constraint will not affect the applicability of the procedure imitate to the satisfiability problem we are interested in.

One key idea of the procedure imitate is that the mimicking process must adhere as much as possible to the original. Since this likeness cannot be realized in the structure of sets, it can be done at least in the cardinality of them. Indeed, the procedure follows the behaviour of the original process keeping equal the cardinality of the regions and the amount of the elements that $\mathcal{P}$-nodes distribute.

When the delivering partition $A_{\mu_i}^{(\mu_i)}$ has not enough elements, μ_i is scant. This will be apparent from the way the sub-procedure revise$_1$ will correlate the increases $\Delta^{[i]}(q)$ made at the i-th step of the mimicking process with the increases $\Delta^{(\mu_i)}(q)$ at the μ_i-th step of the original process. Not only the cardinalities of the $\Delta^{[i]}$-blocks, but also the destination block for the set $\bigcup \widehat{A}_{\mu_i}^{[i]}$, will be constrained in this case to comply with the original.

As soon as we leave the original process, the role of ρ becomes essential in concretely building the shadow process. This use of ρ is strictly related to the sub-procedure revise$_1$. Indeed, the chosen node has at least one block not small, therefore the node has enough elements to propagate ρ new elements to each target and to trigger off the desired "chain reaction".

We stress the fact that, in this context, the concrete possibility to copy a formative process is strictly related to pure cardinality-arguments; there is only one exception: the principal target.

How to choose the destination for the set $\bigcup \widehat{A}_{\mu_i}^{[i]}$ at the i-th step of the mimicking process is, as a matter of fact, another crucial issue in the design of imitate: in fact, when there is a chance that μ_i is the last μ for which $A_{\mu_i}^{(\mu)}$ is the delivering partition (which may happen when μ_i is scant, and certainly is the case when μ_i is a finishing move—recall, in fact, that the original process is greedy), one should be aware that choosing where to place $\bigcup \widehat{A}_{\mu_i}^{[i]}$ possibly affects whether or not certain membership literals will be modeled as desired at the end of the mimicking process. This concern will be apparent both in revise$_1$ and in revise$_3$; slightly less evident in revise$_2$, because of criterion 3 having priority over criterion 2. The occasion in which $\bigcup A_{\mu_i}^{(\bullet)}$ reaches its destination block is in fact the last time A_{μ_i} gets exploited for a step in the original process; hence it is also the last time it will be exploited in the mimicking process, and the way this step will be affected depends on whether μ_i is scant or not. In the latter case the step will fall under the jurisdiction of either revise$_3$ or revise$_2$ depending on whether its destination is a genuine block of $\Pi^{(\bullet)}$ or the fictitious block σ^*. Hence revise$_3$ (very much like revise$_1$) should base its decision about where to place $\bigcup \widehat{A}_{\mu_i}^{[i]}$ on the imitation of the original process; as for revise$_2$, its task is less demanding, since it should simply avoid putting $\bigcup \widehat{A}_{\mu_i}^{[i]}$ in any genuine block of the simulating partition.

At this point we are ready to display in full the above-outlined procedure:

procedure   imitate(ϱ, Π, $(A_\nu)_{\nu<\xi}$);

C_0 : **claim**

$|\Pi| < \aleph_0 \wedge \varrho > L \wedge 2^{\varrho-1} > \varrho \cdot |\Pi| \wedge (A_\mu)_{\mu<\xi}$ is the trace associated greedy formative process $(\{q^{(\mu)}\}_{q\in\Pi})_{\mu\leqslant\xi}$ such that $P^{(\bullet)} \not\subseteq \{\{\emptyset\}\}$;

explanation: the formative process $(\{q^{(\mu)}\}_{q\in\Pi})_{\mu\leqslant\xi}$ will be taken as an oracle in what follows, by referring to the blocks $q^{(\mu)}$ with $q \in \Pi$ and to the partitions $A_\nu^{(\mu)}$ as if they were available as additional inputs;

$M_1 := \left\{ \mu \mid \mu < \xi \wedge (\forall p \in A_\mu)\, \big(|p^{(\mu)}| < \varrho\big) \right\}$;

$M_2 := \left\{ \mu \mid \mu < \xi \wedge (\exists q \in \Pi)\, \left(\big(|q^{(\mu)}| < \varrho \vee q^{(\mu)} \cap \mathcal{P}^*(A_\mu^{(\mu)}) = \emptyset\big) \wedge \Delta^{(\mu)}(q) \neq \emptyset \right) \right\}$;

$M_3 := \{ \mu \mid \mu < \xi \wedge \bigcup A_\mu^{(\mu)} = \bigcup A_\mu^{(\bullet)} \in \bigcup P^{(\bullet)} \}$;

$Sal := M_1 \cup M_2 \cup M_3$;   -- salient steps

$Sc := M_1$;   -- scant (salient) steps

$Fi := M_3 \setminus M_1$;   -- finishing (salient) steps

$In := M_2 \setminus M_1 \setminus M_3$;   -- innovative (salient) steps

A_0 : **assert** $|Sal| < \varrho \cdot |\Pi| \cdot 2^{|\Pi|} + 3$;

let $\{\mu_0, \mu_1, \ldots, \mu_\ell\} = Sal \cup \{\xi\}$, **with** $\mu_0 < \mu_1 < \cdots < \mu_\ell$;

for $q \in \Pi$ **do** $\widehat{q} := \emptyset$; $\Delta(q) := \emptyset$; **end for;**

notation: throughout, and for all $B \subseteq P$, $\widehat{B} := \{\widehat{q} \mid q \in B\}$;

for $i \in [0, \ldots, \ell]$ **do** -- the main loop begins here

for $q \in \Pi$ **do** $\widehat{q} := \widehat{q} \cup \Delta(q)$; **end for;**

C_1 :   **claim**

$(\forall q \in \Pi)\Big(\big((|q^{(\mu_i)}| < \varrho \vee |\widehat{q}| < \varrho) \rightarrow |\widehat{q}| = |q^{(\mu_i)}| \big)$

$\wedge (\forall B \subseteq \Pi)\big((\forall p \in B)(|p^{(\mu_i)}| < \varrho)$

$\rightarrow |q^{(\mu_i)} \cap \mathcal{P}^*(B^{(\mu_i)})| = \big|\widehat{q} \cap \mathcal{P}^*(\widehat{B})\big| \big)$

$\wedge (\forall B \subseteq \Pi)\big(q^{(\mu_i)} \cap \mathcal{P}^*(B^{(\mu_i)}) \neq \emptyset \leftrightarrow \widehat{q} \cap \mathcal{P}^*(\widehat{B}) \neq \emptyset \big) \Big)$;

if $i = \ell$ **then quit for-loop; end if;**

-- the seemingly useless last iteration calls for a final verification of C_1

if $\mu_i \in Sc$ **then** $\Delta := \text{revise}_1(\mu_i, \widehat{\ });$ **end if;**

if $\mu_i \in In$ **then** $\Delta := \text{revise}_2(\mu_i, \widehat{\ });$ **end if;**

if $\mu_i \in Fi$ **then** $\Delta := \text{revise}_3(\mu_i, \widehat{\ });$ **end if;**

C_2 :   **claim**

subPartitions(Δ, A_{μ_i}) $\wedge (\bigcup \widehat{\Pi}) \cap \bigcup \Delta[\Pi] = \emptyset$;

C_3 :   **claim**

$(\forall q \in \Pi)\big(\Delta^{(\mu_i)}(q) \neq \emptyset \Leftrightarrow \Delta(q) \neq \emptyset \big)$;

end for;

C_4 : **claim**

$\widehat{\Pi}$ is a transitive partition ϱ-imitating $\Pi^{(\bullet)}$, hence $\widehat{\Pi}$ L-simulates $\Pi^{(\bullet)}$;

return $\widehat{\Pi}$;

procedure revise$_1$(μ, $\widehat{\ }$);

$A := A_\mu$;

A_1 : **assert**

$\Big(\exists \{\Delta(r)\}_{r\in\Pi} \Big) \Big($

subPartitions(Δ, A) $\wedge$

$(\forall q \in \Pi)\big(|\Delta(q)| = |\Delta^{(\mu)}(q)| \wedge$

$(\bigcup \widehat{A} \in \Delta(q) \leftrightarrow \bigcup A^{(\mu)} \in \Delta^{(\mu)}(q)) \big) \Big)$;

pick one such Δ;   **return** Δ;

end revise$_1$;

procedure revise$_2$(μ, $\widehat{\ }$);

$$A := A_\mu;$$

$\mathsf{A}_2:$ **assert**

$$\Big(\exists\,\{\Delta(p)\}_{p\in\Pi}\Big)\Big(\;\mathsf{subPartitions}(\;\Delta,\,A\;)\,\wedge\,\bigcup\widehat{A}\notin\bigcup\Delta[\Pi]\,\wedge$$

$$\Big(\forall\,q\in\Pi\Big)\Big($$

$$\quad\textbf{if }\big|q^{(\mu+1)}\big|<\varrho\vee\Delta^{(\mu)}(q)=\emptyset\textbf{ then}$$

$$\qquad\big|\Delta(q)\big|=\big|\Delta^{(\mu)}(q)\big|$$

$$\quad\textbf{else if }q^{(\mu)}\cap\mathcal{P}^*(A^{(\mu)})=\emptyset\,\wedge\,|\widehat{q}|\geqslant\varrho\textbf{ then}$$

$$\qquad\big|\Delta(q)\big|\geqslant 1$$

$$\quad\textbf{else}$$

$$\qquad\big|\Delta(q)\big|\geqslant\varrho-|\widehat{q}|$$

$$\quad\textbf{end if }\;\Big)\,\wedge$$

$$\Big|\mathcal{P}^*(\widehat{A})\setminus\bigcup\widehat{\Pi}\setminus\bigcup\Delta[\Pi]\Big|\geqslant 1\;+$$

$$+\sum_{\substack{r\in\Pi\\|r^{(\bullet)}|<\varrho}}\big|\big(r^{(\bullet)}\setminus r^{(\mu+1)}\big)\cap\mathcal{P}^*(A^{(\bullet)})\big|+\sum_{\substack{r\in\Pi\\|r^{(\bullet)}|\geqslant\varrho\\|r^{(\mu+1)}|<\varrho}}\big(\varrho-\big|r^{(\mu+1)}\big|\big)$$

$$+\big|\{r\in\Pi\mid\big|r^{(\mu+1)}\big|\geqslant\varrho\wedge r^{(\bullet)}\cap\mathcal{P}^*(A^{(\bullet)})\neq\emptyset\wedge r^{(\mu+1)}\cap\mathcal{P}^*(A^{(\mu)})=\emptyset\}\big|$$

$$\Big);$$

pick one such Δ; **return** Δ;

end revise$_2$;

procedure revise$_3(\;\mu,\;\widehat{\ }\;)$;

$$A := A_\mu;$$

$\mathsf{A}_3:$ **assert**

$$\Big(\exists\,\{\Delta(p)\}_{p\in\Pi}\Big)\Big(\;\mathsf{subPartitions}(\;\Delta,\,A\;)\,\wedge$$

$$\Big(\forall\,q\in\Pi\Big)\Big(\;\big(\;\bigcup\widehat{A}\in\Delta(q)\;\leftrightarrow\;\bigcup A^{(\bullet)}\in q^{(\bullet)}\;\big)\,\wedge$$

$$\quad\textbf{if }\big|q^{(\mu+1)}\big|<\varrho\vee\Delta^{(\mu)}(q)=\emptyset\textbf{ then}$$

$$\qquad\big|\Delta(q)\big|=\big|\Delta^{(\mu)}(q)\big|$$

$$\quad\textbf{else if }q^{(\mu)}\cap\mathcal{P}^*(A^{(\mu)})=\emptyset\,\wedge\,|\widehat{q}|\geqslant\varrho\textbf{ then}$$

$$\qquad\big|\Delta(q)\big|\geqslant 1$$

$$\quad\textbf{else}$$

$$\qquad\big|\Delta(q)\big|\geqslant\varrho-|\widehat{q}|$$

$$\quad\textbf{end if }\;\Big)\,\wedge$$

$$\big(\;\mathcal{P}^*(A^{(\bullet)})\subseteq\bigcup P^{(\bullet)}\;\rightarrow\;\bigcup\Delta[\Pi]=\mathcal{P}^*(\widehat{A})\setminus\bigcup\widehat{\Pi}\;\big)$$

$$\Big);$$

pick one such Δ; **return** Δ;

end revise$_3$;

procedure subPartitions($\Delta,\,B$);

return

$$\emptyset\neq\bigcup\Delta[\Pi]\subseteq\mathcal{P}^*(\widehat{B})\setminus\bigcup\widehat{\Pi}\,\wedge\,(\forall\,q,r\in\Pi)\big(\,q\neq r\;\rightarrow\;\Delta(q)\cap\Delta(r)=\emptyset\,\big);$$

end subPartitions;

end imitate.

4.2.5   Meaning of Main Claim- and Assert-Statements Occurring Inside imitate

In order to make effective the procedures described in the previous section, it is necessary to show that the requested claims and asserts can be really achieved. We check that all claim- and assert-statements occurring inside the procedure imitate are fulfilled whenever such statements are met during execution, or simply we get a clear overall view of what the procedure does. We refer for the complete details of these proofs to [COU02]. Here, instead, we give some of the main ideas which are fundamental to understand how the analysis must be conducted.

The claims to be proved are C_1–C_4 only: C_0, in fact, expresses conditions which the input parameters are supposed to comply with. The statements A_0–A_3 are not important in themselves, but only because their achievement is essential in order to continue the algorithm. They are not the main purpose of the procedure but it is impossible to guarantee the correct calculation of the output shadow process without their verification at each step. In fact, A_1–A_3 claim the existence of partitioning functions which are referred to by subsequent executable statements.

In other words, there is a conceptual difference between statements and asserts. Indeed while claims show that the result of the procedure is a transitive partition which imitates the original one, the asserts say that this combinatorial mechanism can really be performed. Nonetheless some assert could have some relevant implication which could be, possibly, essential in order to prove some claim. For example, A_3 plays a fundamental role in order to prove that if the starting partition is a $\mathcal{P}$-partition, then the resulting partition is a $\mathcal{P}$-partition as well.

What the procedure simply does is to build step by step the original Π-board. This is in a sense more than it is expected to do, since in order to simulate one does not need to activate all the edges. Nonetheless, as we shall see more in detail in Remark 4.6, we do this for some kind of sense of symmetry and because it can be done without affecting the general result.

More in detail, the third conjunct of claim C_1 is

$$(\forall q \in \Pi)(\forall B \subseteq \Pi)\left(q^{(\mu_i)} \cap \mathcal{P}^*(B^{(\mu_i)}) \neq \emptyset \;\leftrightarrow\; \widehat{q} \cap \mathcal{P}^*(\widehat{B}) \neq \emptyset \right), \tag{4.2}$$

as result of the above request.

We chose to force (4.2), at the price of some complications in the procedures imitate, revise$_2$, and revise$_3$; our gain is that we have set the ground for further applications of the formative process technique to the set-theoretic satisfiability problem as clarified in the following remark.

Remark 4.6. It is worth mentioning that, in general, whenever one is faced with an extended language, which means that it admits new set-constructors/relators, one has also to extend the structural properties meant to be preserved. Rather surprisingly, when we move from MLSSP to the enriched version MLSSPF, we do not need to select more steps, provided that one is not so narrow minded in selecting steps. In other words, the shadow process created in the case of MLSSP-formulae is not the shortest possible but preserves some kind of symmetry. This reveals its power in solving the decision problem for MLSSPF-formulae. In fact, we do not need a new strategy in order to create a suitable shadow process. In particular, such a shadow process automatically reveals objects that one does not need to define for the decision problem of MLSSP, namely, pumping chains (see Chapter 5).

Indeed, the fulfillment of two new conditions (cf. (4.3) and (4.4) below) are exactly what we need for extensions of the fragment of set theory with literals of the forms $\mathrm{Finite}(v)$, $\neg\mathrm{Finite}(v)$, described in the next chapter on MLSSPF.

We need to make more explicit the intended meaning of boards uniformly isomorphic which is the fundamental idea underneath the notion of shadow process. With any given formative process $\mathcal{Q}_{P,\xi} = (\Pi^{(\mu)})_{\mu \leqslant \xi}$, we can associate a sequence $\mathcal{G}_{\mathcal{Q}_{P,\xi}} = (G^{(\mu)})_{\mu \leqslant \xi}$ of labelled directed graphs defined in the following way:

$$\text{nodes}(G^{(\mu)}) = \mathcal{P}(\Pi),$$
$$\text{edges}(G^{(\mu)}) = \{[A, p, B] \mid A, B \subseteq \Pi \wedge p \in B \wedge p^{(\mu)} \cap \mathcal{P}^*(A^{(\mu)}) \neq \emptyset\},$$

for $\mu \leqslant \xi$ (notice that $[A, p, B]$ denotes the edge $[A, B]$ with label p).

Observe that the sequence $\mathcal{G}_{\mathcal{Q}_{P,\xi}}$ is monotone non-decreasing. Let $\mathit{skel}(\mathcal{G}_{\mathcal{Q}_{P,\xi}})$ denote the longest monotone increasing subsequence of $\mathcal{G}_{\mathcal{Q}_{P,\xi}}$.

Next, let $\widehat{\mathcal{Q}}_{P,\ell} = (\widehat{\Pi}^{[j]})_{j \leqslant \ell}$ be a weak formative process generated by the procedure imitate in correspondence of an input formative process $(\Pi^{(\mu)})_{\mu \leqslant \xi}$, and let $\mathcal{G}_{\widehat{\mathcal{Q}}_{P,\ell}} = (\widehat{G}^{[j]})_{j \leqslant \ell}$ be its associated graph sequence.

Then it can be shown that thanks to (4.2) the following two conditions are met:

$$\mathit{skel}(\mathcal{G}_{\mathcal{Q}_{P,\xi}}) = \mathit{skel}(\mathcal{G}_{\widehat{\mathcal{Q}}_{P,\ell}}), \tag{4.3}$$

$$\widehat{G}^{[j]} = G^{(\mu_j)}, \qquad \text{for } 0 \leqslant j \leqslant \ell. \tag{4.4}$$

In this context it could be helpful to see Figure 3.4 in Section 3.4, where there is an immediate and intuitive evidence of this phenomenon.

By using the assert-statements in revise$_1$, it is straightforward to verify that

$$\bigcup B^{(\bullet)} \in \Delta^{(\mu_j)}(q) \;\leftrightarrow\; \bigcup \widehat{B}^{[\ell]} \in \Delta^{[j]}(q), \tag{4.5}$$

for every $B \subseteq P$, $q \in P$, and $0 \leqslant j \leqslant \ell$. This brings us closer to making the output formative process a shadow process of the original process. Indeed, the above property plainly implies Definition 3.21(c). On the other side, C_1 implies 3.21(a). Therefore, in order to prove that the formative process resulting from the procedure imitate is a shadow process of the original one, we are left to prove just property 3.21(b),. In any case these arguments will be extensively treated in Lemma 4.8.

If one is not interested in properties (4.3) and (4.4) above, as is the case when one wants to take into account only the application to the decision problem for Boolean combinations of literals of the following types

$$\begin{array}{llllll} x = y, & x \neq y, & x = \emptyset, & x = y \cup z, & x = y \cap z, \\ x = y \setminus z, & x \subseteq y, & x \not\subseteq y, & x \in y, & x \notin y, \\ x = \mathcal{P}(y), & x = \{y_1, \ldots, y_H\}, & & & \end{array} \tag{4.6}$$

then the following simplifications can be made to the procedure imitate:

- define M_2 as the set $\{\mu \mid \mu < \xi \wedge (\exists q \in \Pi)\,(\,|q^{(\mu)}| < \varrho \wedge \Delta^{(\mu)}(q) \neq \emptyset\,)\}$;

- eliminate the conjunct (4.2) from claim C_1;

- replace revise$_2$ and revise$_3$ by the procedures revise$_2'$ and revise$_3'$ shown in Figure 4.3. ∎

procedure   revise$_2'$(μ, ⌃);
     $A := A_\mu$;

A_2 :   assert

$$\Big(\exists\, \{\Delta(p)\}_{p \in \Pi} \Big) \Big(\; \mathsf{subPartitions}(\; \Delta,\, A\;) \wedge \bigcup \widehat{A} \notin \bigcup \Delta[\Pi] \wedge$$

$$\big(\forall q \in \Pi \big) \Big($$

$$\textbf{if } \big|q^{(\mu+1)}\big| < \varrho \vee \Delta^{(\mu)}(q) = \emptyset \textbf{ then}$$

$$|\Delta(q)| = \big|\Delta^{(\mu)}(q)\big|$$

$$\textbf{else}$$

$$|\Delta(q)| \geqslant \varrho - |\widehat{q}|$$

$$\textbf{end if } \Big) \wedge$$

$$\Big| \mathcal{P}^*(\widehat{A}) \setminus \bigcup \widehat{\Pi} \setminus \bigcup \Delta[\Pi] \Big| \geqslant 1 +$$

$$+ \sum_{\substack{r \in \Pi \\ |r^{(\bullet)}| < \varrho}} \big|(r^{(\bullet)} \setminus r^{(\mu+1)}) \cap \mathcal{P}^*(A^{(\bullet)})\big| + \sum_{\substack{r \in \Pi \\ |r^{(\bullet)}| \geqslant \varrho \\ |r^{(\mu+1)}| < \varrho}} \big(\varrho - |r^{(\mu+1)}| \big)$$

$$\Big);$$

     pick one such Δ;   **return** Δ;
  end   revise$_2'$;

procedure   revise$_3'$(μ, ⌃);
     $A := A_\mu$;

A_3 :   assert

$$\Big(\exists\, \{\Delta(p)\}_{p \in \Pi} \Big) \Big(\; \mathsf{subPartitions}(\; \Delta,\, A\;) \wedge$$

$$\big(\forall q \in \Pi \big) \Big(\Big(\bigcup \widehat{A} \in \Delta(q) \leftrightarrow \bigcup A^{(\bullet)} \in q^{(\bullet)} \Big) \wedge$$

$$\textbf{if } \big|q^{(\mu+1)}\big| < \varrho \vee \Delta^{(\mu)}(q) = \emptyset \textbf{ then}$$

$$|\Delta(q)| = \big|\Delta^{(\mu)}(q)\big|$$

$$\textbf{else}$$

$$|\Delta(q)| \geqslant \varrho - |\widehat{q}|$$

$$\textbf{end if } \Big) \wedge$$

$$\Big(\mathcal{P}^*(A^{(\bullet)}) \subseteq \bigcup P^{(\bullet)} \;\rightarrow\; \bigcup \Delta[\Pi] = \mathcal{P}^*(\widehat{A}) \setminus \bigcup \widehat{\Pi} \Big)$$

$$\Big);$$

     pick one such Δ;   **return** Δ;
  end   revise$_3'$;

Figure 4.3: Simplified variants of revise$_2$ and revise$_3$

4.2.6   Correctness of Procedure imitate

Preliminarily to proving that the process created by the procedure imitate is actually a shadow process, we have to check that it is a formative process.

Lemma 4.7. *Let* $0 \leqslant k \leqslant \ell$. *If none of the statements* C_1–C_3, A_1–A_3 *in* imitate *ever gets violated, for* $i = 0, 1, \ldots, k-1$, *then the functions* $\{\widehat{q}^{[j]}\}_{q \in \Pi}$ ($j = 0, 1, \ldots, k$) *make a weak formative process on* $\{\, q \in \Pi \mid \widehat{q}^{[k]} \neq \emptyset \,\}$, *with trace* $A_{\mu_0}, \ldots, A_{\mu_{k-1}}$.

Proof. Regular termination of the k-th iteration of the main loop of imitate is obviously ensured by the assumption that the assert-statements are fulfilled every time they get reached. Then one observes that $\widehat{q}^{[0]} = \emptyset$ and $\widehat{q}^{[i+1]} = \widehat{q}^{[i]} \cup \Delta^{[i]}(q)$, where $\{\, \Delta^{[i]}(q) \mid q \in \Pi \,\} \setminus \{\emptyset\}$ is a partition of some non-null $Q \subseteq \mathcal{P}^*(\widehat{A}_{\mu_i}^{[i]}) \setminus \bigcup \widehat{\Pi}^{[i]}$, by C_2, as inspection of subPartitions reveals. By contrasting all of this with the definition of formative process, one sees that the thesis holds.   □

We are now ready to prove that the process generated by procedure imitate is a shadow process.

Lemma 4.8. *Procedure* imitate *creates a process which is a shadow process of the input process with synchronization map* $\varsigma \colon \{0, 1, \ldots, \ell\} \twoheadrightarrow \{0, 1, \ldots, \xi\}$. *Moreover, if* $\Pi^{(\bullet)}$ *is a $\mathcal{P}$-partition, then* $\widehat{\Pi}^{[\ell]}$ *partition is a $\mathcal{P}$-partition too.*

Proof. In order to prove our thesis, we have first to verify that $\widehat{\Pi}^{[\ell]}$ is a transitive partition. Thanks to Lemma 4.7 and to Lemma 3.8, we can simplify (i) into $\emptyset \notin \widehat{\Pi}^{[\ell]}$. Our task, accordingly, will be to prove that $\widehat{q}^{[\ell]} \neq \emptyset$ holds for each $q \in \Pi$. Since $q^{(\bullet)} \neq \emptyset$ and $q^{(\bullet)} = \bigcup_{0 \leqslant \mu < \xi} \Delta^{(\mu)}(q)$, it makes sense to consider the least ordinal $\overline{\mu}$ for which $\Delta^{(\overline{\mu})}(q) \neq \emptyset$; i.e., $\Delta^{(\overline{\mu})}(q) \neq \emptyset \wedge (\forall \mu)(0 \leqslant \mu < \overline{\mu} \rightarrow \Delta^{(\mu)}(q) = \emptyset)$, whence $q^{(\overline{\mu})} = \emptyset$ easily follows.

We immediately notice that $\overline{\mu} \in M_2$, so that $\overline{\mu} = \mu_{i_0}$ for some $i_0 < \ell$; thus, if we manage to prove that $\Delta^{[i_0]}(q) \neq \emptyset$, then we can conclude that $\widehat{q}^{[\ell]} \neq \emptyset$, because $\widehat{q}^{[\ell]} \supseteq \Delta^{[i_0]}(q)$. Notice that $\widehat{q}^{[i_0]} = \emptyset$ ensues from $q^{(\mu_{i_0})} = \emptyset$, thanks to the first conjunct in C_1.

If $\overline{\mu} \in Sc$ then, by the assert-statement in revise$_1$, $\left| \Delta^{[i_0]}(q) \right| = \left| \Delta^{(\overline{\mu})}(q) \right|$ holds. If $\overline{\mu} \in Fi \cup In$, then inspection of the assert-statements A_3 and A_2 reveals two possibilities only: either $\left| \Delta^{[i_0]}(q) \right| = \left| \Delta^{(\overline{\mu})}(q) \right|$, or $\left| \Delta^{[i_0]}(q) \right| \geqslant \varrho - \left| \widehat{q}^{[i_0]} \right| = \varrho > 0$. In either case $\Delta^{[i_0]}(q) \neq \emptyset$, and hence our thesis $\widehat{q}^{[\ell]} \neq \emptyset$ holds.

Now we verify that the shadow conditions hold and that $\widehat{\Pi}^{[\ell]}$ is a $\mathcal{P}$-partition whenever $\Pi^{(\bullet)}$ is also a $\mathcal{P}$-partition.

(a) (Edge Activation) $\mu_{\widehat{\mathsf{EA}(B,q)}} = \mathsf{EA}(B, q)$, for every $B \subseteq \Pi$ and $q \in \Pi$.

(b) (Distribution) $\Delta^{[i]}(q) \neq \emptyset \Longleftrightarrow \Delta^{(\mu_i)}(q) \neq \emptyset$, for every $q \in \Pi$ and $i \in \mathrm{dom}(\varsigma)$.

(c) (Last Distribution) For every node $B \subseteq \Pi$, $\mu_{\widehat{\mathsf{GE}(B)}} = \mathsf{GE}(B)$. Moreover, $\bigcup \widehat{B} \in \Delta^{[\widehat{\mathsf{GE}(B)}]}(q) \leftrightarrow \bigcup B^{(\bullet)} \in \Delta^{(\mathsf{GE}(B))} q^{(\bullet)}$.

(d) If $\Pi^{(\bullet)}$ is a $\mathcal{P}$-partition then $\widehat{\Pi}^{[\ell]}$ is a $\mathcal{P}$-partition.

(a):   This property plainly follows from the distribution property and the fact that if in the original process $\mu = \mathsf{EA}(B, q)$, then μ is a salient step (cf. Exercise 4.2).

(b):   This is basically the same of Claim C_3.

(c):   If $\mu = \mathsf{GE}(B)$, then it can be either scant or finishing; therefore $\mu = \mu_i$ for some i. In both cases, just by checking assert A_1 or A_3, it can be seen that $\bigcup \widehat{A} \in \Delta(q) \leftrightarrow \bigcup A^{(\mu)} \in \Delta^{(\mu)}(q)$. What is left to prove is that $i = GE(\widehat{B})$ or, in other terms, that for all $\widehat{q} \in \widehat{B}$ and, for all $j > i$, $\Delta^{[j]}(\widehat{q}) = \emptyset$. But this follows from Claim C_3 and the fact that $\mu_i = \mathsf{GE}(B)$. On the other side, if $i = \mathsf{GE}(\widehat{B})$, then by assert A_2 $\mu_i \notin In$, so that either $\mu_i \in Sc$ or $\mu_i \in Fi$. In both cases $\bigcup \widehat{A} \in \Delta(q) \leftrightarrow \bigcup A^{(\mu)} \in \Delta^{(\mu)}(q)$. Arguing as in the first part of the proof we can conclude that μ_i is equal to $\mathsf{GE}(B)$.

(d):   It is sufficient to prove that if $\mathcal{P}^*(B^{(\bullet)}) \subseteq \bigcup \Pi^{(\bullet)}$, then $\mathcal{P}^*(\widehat{B}^{[\ell]}) \subseteq \bigcup \widehat{\Pi}^{[\ell]}$. The reader is asked to prove in the exercises that the previous property implies (d). (This proof is the same as that of [COU02, Lemma 8.2(iv)]; we report it here for the reader's convenience.) Let $\mathcal{P}^*(B^{(\bullet)}) \subseteq \bigcup \Pi^{(\bullet)}$. Then, since $\bigcup B^{(\bullet)} \in \mathcal{P}^*(B^{(\bullet)}) \subseteq \bigcup \Pi^{(\bullet)}$, where

$\bigcup \Pi^{(\bullet)} = \bigcup_{p \in \Pi} \bigcup_{\mu < \xi} \Delta^{(\mu)}(p)$, we have $\bigcup B^{(\bullet)} \in \Delta^{(\overline{\mu})}(p)$ for suitable $p \in \Pi$ and $\overline{\mu} < \xi$. Hence $B = A_{\overline{\mu}}$ and $\bigcup A_{\overline{\mu}}^{(\overline{\mu})} = \bigcup A_{\overline{\mu}}^{(\bullet)}$, and therefore $\overline{\mu} \in M_3$. Let $i_0 < \ell$ be such that $\overline{\mu} = \mu_{i_0}$.

From $\bigcup A_{\overline{\mu}}^{(\overline{\mu})} = \bigcup A_{\overline{\mu}}^{(\bullet)}$ it follows that $\Delta^{(\mu)}(p) = \emptyset$, for $p \in A_{\overline{\mu}}$ and $\overline{\mu} \leqslant \mu < \xi$, whence, by claim C_3, $\Delta^{[i]}(p) = \emptyset$ ensues for $p \in A_{\overline{\mu}}$ and $i_0 \leqslant i < \ell$. Hence $\widehat{A}_{\overline{\mu}}^{[i_0]} = \widehat{A}_{\overline{\mu}}^{[\ell]}$.

Our goal in what follows is to show that

$$\bigcup \Delta^{[i_0]}[\Pi] = \mathcal{P}^*(\widehat{A}_{\overline{\mu}}^{[i_0]}) \setminus \bigcup \widehat{\Pi}^{[i_0]} \tag{4.7}$$

holds; this will readily yield that

$$\mathcal{P}^*(\widehat{A}_{\overline{\mu}}^{[\ell]}) \;=\; \mathcal{P}^*(\widehat{A}_{\overline{\mu}}^{[i_0]}) \;\subseteq\; \bigcup \widehat{\Pi}^{[i_0]} \;\cup\; \bigcup \Delta^{[i_0]}[\Pi]$$
$$= \bigcup \widehat{\Pi}^{[i_0+1]} \;\subseteq\; \bigcup \widehat{\Pi}^{[\ell]},$$

which encompasses our desired conclusion.

If $\overline{\mu} \in Fi$, then (4.7) follows immediately from the assertion A_3.

On the other hand, if $\overline{\mu} \in Sc$, then we have

- $|\bigcup \Delta^{[j]}[\Pi]| = |\bigcup \Delta^{(\mu_j)}[\Pi]|$ for all j such that $0 \leqslant j \leqslant i_0$ and $A_{\mu_j} = A_{\overline{\mu}}$ (by the assertion A_1);

- $\left| \mathcal{P}^*(\widehat{A}_{\overline{\mu}}^{[i_0]}) \right| = \left| \mathcal{P}^*(A_{\overline{\mu}}^{(\overline{\mu})}) \right|$ (by the statement C_1);

- $\left| \mathcal{P}^*(\widehat{A}_{\overline{\mu}}^{[i_0]}) \cap \bigcup \widehat{\Pi}^{[i_0]} \right| = \sum_{\substack{0 \leqslant j < i_0 \\ A_{\mu_j} = A_{\overline{\mu}}}} |\bigcup \Delta^{[j]}[\Pi]|$

  $\qquad\qquad\qquad\qquad = \sum_{\substack{0 \leqslant j < i_0 \\ A_{\mu_j} = A_{\overline{\mu}}}} |\bigcup \Delta^{(\mu_j)}[\Pi]| = \left| \mathcal{P}^*(A_{\overline{\mu}}^{(\overline{\mu})}) \cap \bigcup \Pi^{(\overline{\mu})} \right|$;

- $\bigcup \Delta^{(\overline{\mu})}[\Pi] = \mathcal{P}^*(A_{\overline{\mu}}^{(\overline{\mu})}) \setminus \bigcup \Pi^{(\overline{\mu})}$

  (by $\mathcal{P}^*(A_{\overline{\mu}}^{(\bullet)}) \subseteq \bigcup \Pi^{(\bullet)}$ and the greediness requirement).

Therefore

$$|\bigcup \Delta^{[i_0]}[\Pi]| \;=\; |\bigcup \Delta^{(\overline{\mu})}[\Pi]| = \left| \mathcal{P}^*(A_{\overline{\mu}}^{(\overline{\mu})}) \setminus \bigcup \Pi^{(\overline{\mu})} \right|$$
$$= \left| \mathcal{P}^*(A_{\overline{\mu}}^{(\overline{\mu})}) \right| - \left| \mathcal{P}^*(A_{\overline{\mu}}^{(\overline{\mu})}) \cap \bigcup \Pi^{(\overline{\mu})} \right|$$
$$= \left| \mathcal{P}^*(\widehat{A}_{\overline{\mu}}^{[i_0]}) \right| - \left| \mathcal{P}^*(\widehat{A}_{\overline{\mu}}^{[i_0]}) \cap \bigcup \widehat{\Pi}^{[i_0]} \right|$$
$$= \left| \mathcal{P}^*(\widehat{A}_{\overline{\mu}}^{[i_0]}) \setminus \bigcup \widehat{\Pi}^{[i_0]} \right|$$

holds, which in turn plainly implies (4.7), concluding our proof. □

We refer to the process generated by procedure imitate as "shadow process-imitate".

Now we discuss the core claim C_4 that actually is what procedure imitate is made for. In order to check C_4, arguing from Lemma 4.8, we are left to verify just Definition 2.26 (iv) as the following lemma does:

Lemma 4.9. $\widehat{\Pi}^{[\ell]}$ L-imitates $\Pi^{(\bullet)}$.

Proof. According to Lemma 4.8 and Definition 2.26, in order to prove the claim we are left to verify (iv): $|\beta(\sigma)| = |\sigma|$ when $|\sigma| < \varrho$. Observe that $\varrho > L$.

Consider the last step μ which gives elements to σ, therefore $\mu \in M_1 \cup M_2 \cup M_3$. By the first conjunct of the claim C_1, we have $|\widehat{\sigma}| = |\sigma^{(\mu_i = \mu)}|$. Since μ is the last step which increases σ, $\widehat{\sigma}^{[i]} = \widehat{\sigma}^{[\ell]}$; thus, by $|\widehat{\sigma}^{[\ell]}| = |\sigma^{(\bullet)}|$, the thesis follows. □

4.2.7   Rough Assessment of the Complexity of imitate

The following lemma not only shows that the cardinality $|Sal|$ is finite (thereby ensuring that the main loop of imitate will be executed finitely many times), but even tightens w.r.t. A_0 the upper bound on this cardinality, setting the ground for the complexity analysis that will be carried out in Section 4.2.8.

Lemma 4.10. *Assuming the conditions in claim* C_0 *to hold, let* $n = |\Pi|$. *Then*
$$|Sal| \leqslant \varrho \cdot n \cdot 2^{n-1} + 2^{n+1} + (\varrho - 1) \cdot n - 1 < \varrho \cdot n \cdot 2^n + 3.$$

Proof. We begin by first estimating $|Sc|$. Let $B \subseteq P$. Notice that if $A_\mu = A_{\mu'} = B$, with $\mu < \mu'$, $\mu, \mu' \in Sc$, then by the pigeon-hole principle the following inequalities hold:
$$|B| \leqslant \left|\bigcup B^{(\mu)}\right| < \left|\bigcup B^{(\mu')}\right| \leqslant (\varrho - 1)|B| . \text{ Hence,}$$
$$\left|\{\mu \in Sc \mid A_\mu = B\}\right| \leqslant (\varrho - 1)|B| - |B| + 1 = (\varrho - 2)|B| + 1 .$$

Therefore

$$
\begin{aligned}
|Sc| &\leqslant \sum_{B \subseteq P} ((\varrho - 2)|B| + 1) &&= \sum_{i=0}^{n} \binom{n}{i}((\varrho - 2)i + 1) \\
&= (\varrho - 2)\sum_{i=0}^{n} i\binom{n}{i} + 2^n &&= (\varrho - 2)\sum_{i=1}^{n} n\binom{n-1}{i-1} + 2^n \\
&= (\varrho - 2)n\sum_{i=0}^{n-1}\binom{n-1}{i} + 2^n &&= (\varrho - 2)n2^{n-1} + 2^n .
\end{aligned}
$$

In order to make an estimate of $|M_2 \setminus Sc|$, let us put

$$M_2' = \left\{ \mu \mid \mu < \xi \wedge (\exists q \in \Pi)\left(\left|q^{(\mu)}\right| < \varrho \wedge \Delta^{(\mu)}(q) \neq \emptyset\right) \right\}$$

and

$$M_2'' = \left\{ \mu \mid \mu < \xi \wedge (\exists q \in \Pi)\left(q^{(\mu)} \cap \mathcal{P}^*(A_\mu^{(\mu)}) = \emptyset \wedge \Delta^{(\mu)}(q) \neq \emptyset\right) \right\} .$$

Plainly, $M_2 = M_2' \cup M_2''$. We first estimate $|M_2' \setminus Sc|$. Thus, let

$$\Phi(\mu) = \sum_{\left|q^{(\mu)}\right| < \varrho} \left(\varrho - \left|q^{(\mu)}\right|\right)$$

and observe that

- $\Phi(\mu) > \Phi(\mu + 1)$, for all $\mu \in M_2'$,

- $\Phi(\mu) \geqslant \Phi(\mu + 1)$, for all $\mu < \xi$,

- $0 \leqslant \Phi(\mu) \leqslant \varrho n$, for all $\mu < \xi$, and

- $0 \in Sc \cap M_2'$.

From these we immediately get $|M_2' \setminus Sc| \leqslant \varrho n$.
    Next, we estimate $|M_2'' \setminus Sc|$. Let us put

$$M_2''(B, q) := \{\mu \in M_2'' \setminus Sc \mid A_\mu = B \wedge q^{(\mu)} \cap \mathcal{P}^*(B^{(\mu)}) = \emptyset \wedge \Delta^{(\mu)}(q) \neq \emptyset\},$$

for $B \subseteq \Pi$ and $q \in \Pi$, and observe that

- $M_2''(\emptyset, q) = \emptyset$, for all $q \in \Pi$, since $\emptyset = A_\mu$ iff $\mu = 0$ and moreover $0 \in Sc$,

- $|M_2''(B, q)| \leq 1$, for all $B \subseteq \Pi$ and $q \in \Pi$, and

- $M_2'' \setminus Sc \subseteq \bigcup_{\substack{B \subseteq \Pi \\ q \in \Pi}} M_2''(B, q)$.

From these, we immediately obtain $|M_2'' \setminus Sc| \leq n(2^n - 1)$. Therefore

$$|M_2 \setminus Sc| \leq |M_2' \setminus Sc| + |M_2'' \setminus Sc| \leq n(2^n - 1) + \varrho n.$$

Finally we estimate $|Fi|$. Since $0 \in Sc \cap M_3$ and $|\{\mu \in M_3 \mid A_\mu = B\}| \leq 1$, for all $B \subseteq P$, we obtain at once $|Fi| \leq 2^n - 1$.

Summing up, we have

$$|Sal| \leq \varrho n 2^{n-1} + 2^{n+1} + (\varrho - 1)n - 1.$$

Since $n \geq 1$ and $\varrho \geq 3$ (the latter follows from the assumptions $\varrho > L$ and $2^{\varrho-1} > \varrho n$ in claim C_0), an easy inductive argument shows that $\varrho n 2^{n-1} + 2^{n+1} + (\varrho-1)n - 1 < \varrho n 2^n + 3$, thus completing the proof of the lemma. $\square$

Proof of Lemma 4.5. Arguing from Lemma 4.9, it is sufficient to substitute ρ with ρ_0 calculated in Section 4.4. $\square$

4.2.8 Decidability of MLSSP

Now we have in our hands all the tools we need to face the decision problem for MLSSP-formulae. Consider any satisfiable MLSSP-formula Φ with m distinct variables with model M and Venn regions $\Pi^{(\bullet)}(|\Pi| \leq 2^m - 1)$. Consider the Venn partition of the above assignment. There is a unique way to obtain from it a $\mathcal{P}$-partition, therefore, without loss of generality, we can assume that $\Pi^{(\bullet)}$ is such a $\mathcal{P}$-partition. Let $(\Pi^{(\mu)})_{\mu \leq \xi}$ be a formative process for $\Pi^{(\bullet)} = \Pi^{(\xi)}$, whose existence is ensured by the Trace Theorem 3.11.

Let $\mathrm{Vars}(\Phi)$ be the collection of variables occurring in Φ and L the maximum integer such that a literal of the form $v = \{w_0, w_1, \ldots, w_L\}$ appears in Φ.

Since the formula Φ is an MLSSP-formula, all the literals inside it are of the type considered by Lemma 2.24. By Lemma 4.5 procedure imitate creates a transitive partition $\widehat{\Pi}^{[\ell]}$ which L-imitates the original one.

Lemma 2.24 applies and $\widehat{\Pi}^{[\ell]}/\mathfrak{I} \models \Phi$. Lemma 4.5 asserts that the rank of the new model cannot exceed $\max\left(\lceil \frac{25}{24} \cdot \lceil \log |\Pi| \rceil + 5 \rceil, L + 1\right) \cdot |\Pi| \cdot 2^{|\Pi|} + 3$.

From the definition of $\widehat{M}$, we also get $\widehat{M}(v) \subseteq \bigcup \widehat{\Pi}^{[\ell]}$, for every v in $\mathrm{Vars}(\Phi)$. Therefore, since $\widehat{\Pi}^{[\ell]}$ is finite and

$$\mathrm{rk}\, \widehat{\Pi}^{[\ell]} \leq \max\left(\lceil \tfrac{25}{24} \lceil \log |\Pi| \rceil + 5 \rceil, L+1\right) |\Pi|\, 2^{|\Pi|} + 3$$
$$\leq \lceil \tfrac{25}{24} m + 5 \rceil\, 2^{2^m + m} + 3,$$

we obtain that for all v in $\mathrm{Vars}(\Phi)$

$$\mathrm{rk}\, \widehat{M}(v) \leq \mathrm{rk}\, \widehat{\Pi}^{[\ell]} \leq \left\lceil \frac{25}{24} m + 5 \right\rceil 2^{2^m + m} + 3.$$

The preceding discussion plainly entails the following result.

Theorem 4.11 (Decidability of MLSSP). *Let Φ be an MLSSP-formula, and $\mathrm{Vars}(\Phi)$ the collection of variables occurring in it. If Φ is satisfiable, then it is satisfied by a set-valued assignment $\widehat{M}$ such that*

$$\mathrm{rk}\ \widehat{M}[\mathrm{Vars}(\Phi)] \leqslant \left\lceil \frac{25}{24} \cdot m + 5 \right\rceil \cdot 2^{2^m + m} + 3,$$

where $m = |\mathrm{Vars}(\Phi)|$.

Hence, the satisfiability problem for the fragment of set theory consisting of propositional combinations of literals of the form (4.6) is decidable.

EXERCISES

Exercise 4.1. *Let $\Phi \stackrel{\mathrm{Def}}{:=} x = \mathcal{P}(y) \wedge x = z \cup w \wedge w = \{y\} \wedge z \neq \emptyset$. If Φ is satisfiable, find a model M for it, the corresponding $\Sigma^{\mathcal{P}}_M$-board, and a formative process for Σ_M.*

Exercise 4.2. *Prove Lemma 4.8(a).*

Exercise 4.3. *In Lemma 4.8, show that if $\Pi^{(\bullet)}$ is a $\mathcal{P}$-partition and property*

$$\mathcal{P}^*(B^{(\bullet)}) \subseteq \bigcup \Pi^{(\bullet)} \Rightarrow \mathcal{P}^*(\widehat{B}^{[\ell]}) \subseteq \bigcup \widehat{\Pi}^{[\ell]}$$

holds, then $\widehat{\Pi}^{[\ell]}$ is a $\mathcal{P}$-partition.

Exercise 4.4. *Let Φ be an MLSSP-formula, and M a model for it. Assume that the $\Sigma^{\mathcal{P}}_M$-board has no cycles. Find a bound to the rank of the model using a method other than that used in the present chapter.*

Exercise 4.5. *The notion of formative process is a pure abstract one, in the sense that it does not depend on the objects or constructor to which it applies. The following example clarifies this point.*

Instead of sets, consider natural numbers, and instead of the operator $\mathcal{P}^*$ consider the operator S^*, defined much as $\mathcal{P}^*$ in the following manner (but on the nodes of a partition of natural numbers):

- $S^*(\emptyset) = \{1\}$

- $S^*(\{s_1, \ldots s_n\}) = \{x_1 + \cdots + x_n \mid x_1 \in q_1, \ldots, x_n \in q_n\}$.

Thus, for instance, $S^*(\{1, 2\}) = \{2, 4, 3\}$ and $S^*(\{1\}, \{2\}) = \{3\}$. We refer to this framework as (ω, S^*), as opposed to the framework $(\mathcal{V}, \mathcal{P}^*)$ adopted throughout the book.

It is clear that the root of the partition is the place containing the number 1.

Find a Π-board which generates an infinite arithmetic progression of ratio (i.e., common difference) 2 in the setting (ω, S^*).

Chapter 5

Decidability of MLSSPF

The analysis of the decision problem for MLSSPF is rather different from that for MLSSP. First of all, an MLSSPF-formula can contain an explicit literal that forces the model to be infinite (e.g., $\neg Finite(x)$), therefore MLSSPF cannot enjoy the small model property. The second different aspect of this application is that we shall not look for any particular shadow process since we use the same process of the previous application. Indeed, instead of creating a process with the desired structural properties, we shall show that the process we already have in our hands has the expected properties. In other words, rather surprisingly, the process built to solve the decision problem for MLSSP preserves more structural properties than is necessary for the previous application, as we already observed in Remark 4.6. In particular, these properties are sufficient to solve the decision problem for MLSSPF.

It is worth mentioning that the shadow technique is applied in a completely different context. Indeed, procedure ExtendFormativeProcess creates shadow models larger than the starting one instead of smaller, as in the application of the previous chapter.

Throughout this chapter, as in the previous one, the partitions resulting from a formative process are to be considered $\mathcal{P}$-partitions.

5.1 Highlights of a Decision Test for MLSSPF

As remarked above, in the case of MLSSPF we cannot set out a small model property for it, since there are satisfiable MLSSPF-formulae admitting models of infinite rank only (e.g., $\neg Finite(x)$). However, we shall be able to prove for MLSSPF a variant of the small model property, i.e., the SMALL WITNESS-MODEL PROPERTY, which will still yield the decidability of the satisfiability problem for MLSSPF.

To be more specific, given a normalized MLSSPF-conjunction Φ, we denote by Φ^- the formula obtained by dropping from Φ all the literals of any of the forms $Finite(x)$ and $\neg Finite(x)$. Notice that Φ^- is an MLSSP-formula. Therefore in a sense MLSSPF is not a real extension of MLSSP but just a sort of refining of MLSSP. Indeed, MLSSPF detects not only if a formula of MLSSP is satisfiable but also which variables inside the formula can be made infinite and this information lies inside the MLSSP formula. The most intriguing fact is that this information can be grasped just by looking at the structure of the Π-board and the formative process of an MLSSP formula. Actually we have found even more. Indeed, the information, which guarantees that a variable can be made infinite, is inside the same collection of models discovered in the previous chapter, that is models for Φ^- of rank not exceeding $c(|V_\Phi|) = \max\left(\left\lceil \frac{25}{24} \cdot \lceil \log_2 |\Sigma| \rceil + 5 \right\rceil, L + 1\right) \cdot |\Sigma| \cdot 2^{|\Sigma|} + 3$.

The information we speak about allows one to extend a model for Φ^- by way of a certain *pumping mechanism*. Since it can be effectively checked whether a given model M for Φ^- over Vars(Φ), of finite rank, can be 'pumped' to a model M' of Φ, the decidability of the satisfiability problem for MLSSPF follows. We refer to a 'small' model M for Φ^- that can be pumped to a model for Φ as a 'small witness model'.

Roughly speaking, to check whether a given model M for Φ^- can be pumped in such a way that all the literals of the form $\neg Finite(x)$ in Φ are also modeled correctly, we superimpose to the Venn partition Σ_M of M a suitable finite graph structure $\mathcal{G}_M$ (*colored syllogistic board*) and then check whether $\mathcal{G}_M$ contains paths (*pumping chains*), satisfying certain 'dynamic' combinatorial conditions related to a formative process for the partition Σ_M, that cover at least one block for each set Mx that needs to become infinite while it involves no block for any set Mv that has to remain finite.[1] In other words, we simply color by red the regions which cannot be made infinite, and by green the remaining others.

Thus, in first approximation, our satisfiability test for MLSSPF will take the form

> **procedure** FirstSatTestMLSSPF(Φ);
> **if** Φ^- is unsatisfiable **then**
>   **return** "Φ is unsatisfiable";
> **else**
>   **for each** model M for Φ^- whose rank is bounded by $c(|V_\Phi|)$ **do**
>   **if** M can be 'pumped' to a model M' for Φ **then**
>   **return** "Φ is satisfiable";
>   **end if**;
>   **end for**;
>   **return** "Φ is unsatisfiable";
> **end if**;
> **end procedure**;

The fact must be stressed that for a given MLSSPF-conjunction Φ and model M for Φ^-, the pumping mechanism must only make infinite those sets that need to be so, without disrupting any of the already established relationships among sets which ensure the satisfiability by M of all the conjuncts of Φ^-. This is achieved as long as the pumping mechanism transforms the blocks of the Venn partition Σ_M in such a way that the new transitive partition $\Sigma_{M'}$ L-simulates the previous one.

If we further assume that the formula Φ contains also literals of the form $Finite(y)$ and $\neg Finite(y)$, then M' satisfies Φ if its derived partition $\Sigma_{M'}$ not only L-simulates the original one, but also satisfies the following condition:

Condition 1: $\beta(\sigma)$ is finite if and only if σ is finite, for $\sigma \in \Sigma_M$.

5.2   Historical Profile of an Infinite Block

Let $\langle \mathcal{T}, \mathcal{R} \rangle$ be a colored Π-board: what guarantee do we have that a colored Π-process $\langle (\Pi_\mu)_{\mu \leqslant \xi}, (\bullet), \mathcal{T}, \mathcal{R} \rangle$ exists? *A priori*, according to our definitions, none; as a matter of

---

[1] This is the test in the inner **if-then** instruction in procedure FirstSatTestMLSSPF.

fact, should some place have no afferent edges, that is, should there be some $s \in \Pi$ such that $s \notin \mathcal{T}(A)$ for any $A \subseteq \Pi$, then no move A_ν would ever be allowed to bring elements into $s^{(\nu+1)}$, and consequently $s^{(\bullet)}$ could not be a block in a partition.

In view of Corollary 3.12, the situation is quite different if the colored Π-board with which we start is induced by a transitive partition Σ.

Observe that Φ^- is an MLSSP formula. Then, if it is satisfied by a model M which induces a particular colored board, we can apply the previous theorem and obtain a formative process for Σ_M. Then Lemma 4.5 applies, and a new 'small' transitive partition $\widehat{\Sigma}$ is generated which induces a new particular colored board that L-simulates Σ and has rank

$$\mathrm{rk}\,\widehat{\Sigma} \leqslant \max\left(\left\lceil \frac{25}{24} \cdot \lceil \log_2 |\Sigma| \rceil + 5 \right\rceil, L+1\right) \cdot |\Sigma| \cdot 2^{|\Sigma|} + 3$$

(so that MLSSP reflects over the hereditarily finite sets). Instead of being obtained directly from Σ, this $\widehat{\Sigma}$ is obtained through a (weak) formative process $\left(\widehat{\Pi}_j\right)_{j \leqslant \ell}$ which 'mimics' a greedy colored process $\left(\Pi_\mu\right)_{\mu \leqslant \xi}$ corresponding to Σ as stated in the Trace Theorem, and by then putting $\widehat{\Sigma} = \widehat{\Pi}_\ell$. Here $\ell < \aleph_0$. Moreover, by Lemma 4.8 this process is actually a shadow one; therefore, the sequence of edge-activation moves and grand-event moves are 'synchronized' between the two processes. In what follows, we shall refer to $\left(\Pi_\mu\right)_{\mu \leqslant \xi}$ and to $\left(\widehat{\Pi}_j\right)_{j \leqslant \ell}$, as described in Section 4.2.1, respectively, as the *original process* and *shadow process*.

From a general point of view, the way in which the small partition $\widehat{\Sigma}$ portrays Σ is unfaithful to the original: all of its blocks have in fact finite cardinality, regardless of the cardinality of the corresponding blocks in the original partition. Although we have found a finite representation of causes which can make a block in Σ infinite, these causes are to some extent static, viz. features of the Σ-board, and to some other extent dynamic, viz. *'pumping' events* that activate certain special paths of the Σ-board during the original process.

Remark 5.1. On the basis of a careful analysis of how infinity can enter into play in the original process, we shall discover how, starting from the shadow process created as in the previous section, it is possible to infinitely expand the transitive partition obtained through the shadow process without disrupting the capability of L-simulating the original one. More precisely, we infinitely expand the green blocks of our small partition keeping effective the property of L-simulating the original partition, taking advantage of two facts:

1. The sequence of boards induced by Π_{μ_i} and $\widehat{\Pi}_i$ are uniformly isomorphic, for $0 \leqslant i \leqslant \ell$ (see Example 3.22 and Remark 4.6);

2. We can synchronize with each pumping event ν of the original process a regular chain of actions, that can safely be interpolated to the move of the shadow process corresponding to the original ν-th move. ∎

5.3 Infinite and Potentially Infinite Venn Regions

The formative process technique allows one to describe an infinite place as a finite object, more precisely as a cycle in the Π-board.

This is the key fact in order to give a finite representation of an infinite model, which in turn allows one to solve positively the decision problem for MLSSPF.

Observe that (see Remark 5.1) if the original process possesses a cycle, so does the shadow process. Therefore it seems there is nothing more to do. Unfortunately, this is not true. A cycle is, in a sense, a static notion since a formative process is not needed to describe a cycle. Indeed, in order to define a cycle it is not required to say how this cycle was born, namely, we do not need the "history" of the cycle. We describe the structure as the matter of the fact, in other terms when it is already done, no matter how. Another approach would be to describe a cycle as it arises, step by step, from the void throughout the process (see again Example 3.22). This is a dynamical description of a cycle.

The following example shows that a static definition of a cycle is not enough to mimic the infinite production of elements. In the rest of the chapter, we shall show instead that a dynamical description of a cycle does the job. Unfortunately, it is not immediately apparent that the shadow process preserves such a kind of dynamical object. To show this, we should make a considerably bigger effort.

Example 5.2. Consider the formative process

$$
\begin{pmatrix} \alpha^{(\mu)} \\ \beta^{(\mu)} \\ \gamma^{(\mu)} \end{pmatrix}_{\mu \leq 4} = \begin{pmatrix} \emptyset & \{\emptyset\} & \{\emptyset\} & \{\emptyset\} & \{\emptyset, \{\{\emptyset\}, \{\emptyset, \{\emptyset\}\}\} \\ \emptyset & \emptyset & \{\{\emptyset\}\} & \{\{\emptyset\}\} & \{\{\emptyset\}\} \\ \emptyset & \emptyset & \emptyset & \{\{\emptyset, \{\emptyset\}\}\} & \{\{\emptyset, \{\emptyset\}\}\} \end{pmatrix},
$$

with history

$$
\begin{pmatrix} A_\mu \\ T_\mu \end{pmatrix}_{\mu < 4} = \begin{pmatrix} \emptyset & \{\alpha\} & \{\alpha, \beta\} & \{\gamma, \beta\} \\ \{\alpha\} & \{\beta\} & \{\gamma\} & \{\alpha\} \end{pmatrix},
$$

and set

$$
\Phi \overset{\text{Def}}{:=} y \in x.
$$

The assignment

$$
Mx := \alpha \cup \beta, \qquad My := \beta \cup \gamma
$$

satisfies Φ.

Next, consider the formula

$$
\Phi' \overset{\text{Def}}{:=} y \in x \wedge \neg Finite(y).
$$

Observe that the Π-board of the model of Φ contains the cycle

$$
\mathcal{C} := \alpha, A_1, \beta, A_2, \gamma, A_3, \alpha
$$

depitcted in the following picture:

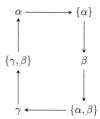

If we try to put infinite elements inside β in order to satisfy $\neg Finite(y)$ by repeating infinitely many times $\mathcal{C}$, we obtain an assignment that satisfies $\neg Finite(y)$, but not $y \in x$ anymore, therefore it does not satisfy Φ'. ∎

We can argue that, provided one changes the formative process, one can anyway satisfy the formula. This is exactly the case of the previous example, although it is not always so. This means that the simple notion of cycle is not enough for our scope, as the following example shows.

Example 5.3. Consider the formative process

$$
\begin{pmatrix} \alpha^{(\mu)} \\ \beta^{(\mu)} \\ \gamma^{(\mu)} \end{pmatrix}_{\mu \leq 4} = \begin{pmatrix} \emptyset & \{\emptyset\} & \{\emptyset\} & \{\emptyset\} & \{\emptyset\}, \{\{\{\{\emptyset\}\}\}\} \\ \emptyset & \emptyset & \{\{\emptyset\}\} & \{\{\emptyset\}\} & \{\{\emptyset\}\} \\ \emptyset & \emptyset & \emptyset & \{\{\{\emptyset\}\}\} & \{\{\{\emptyset\}\}\} \end{pmatrix},
$$

with history

$$
\begin{pmatrix} A_\mu \\ T_\mu \end{pmatrix}_{\mu < 4} = \begin{pmatrix} \emptyset & \{\alpha\} & \{\beta\} & \{\gamma\} \\ \{\alpha\} & \{\beta\} & \{\gamma\} & \{\alpha\} \end{pmatrix},
$$

and let

$$
\Phi \stackrel{\text{Def}}{:=} z = \{y\} \wedge z \in x.
$$

The assignment

$$
Mx := \alpha^4, \quad My := \beta^4, \quad Mz := \gamma^4
$$

satisfies Φ.

However, the formula

$$
\Phi' \stackrel{\text{Def}}{:=} z = \{y\} \wedge z \in x \wedge \neg Finite(z)
$$

is clearly unsatisfiable, even if Φ is satisfiable and z has the cycle

$$
\mathcal{C} := \alpha, A_1, \beta, A_2, \gamma, A_3, \alpha
$$

at γ, as in the next picture.

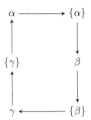

5.3.1 Pumping Paths

The block at a place s cannot become infinite, during a formative process, without some afferent edge carrying to s infinitely many elements drawn from a cycle. Otherwise stated, to make it possible that $\left|s^{(\bullet)}\right| \geqslant \aleph_0$, the board of the process must have a path to s, beginning with a cycle, of the form

$$W_1, \ W_2, \ \ldots, \ W_n \, (= W_1), \ W_{n+1}, \ \ldots, \ W_m \, (= s),$$

enjoying also some additional properties to be stated soon, which will be called a *pumping path* to s. This fact should be intuitively easy to grasp, and we shall provide a proof of it soon. Notice that we can insist w.l.o.g. that, besides the double occurrence of $W_1 = W_n$, no further repetitions of vertices are allowed in the path.

In a colored Π-board $\mathcal{G} = \langle \mathcal{T}, \mathcal{R} \rangle$, a PATH (of length $k - 1$) is any ordered vertex list $W_1, \ldots, W_k$ within which places and nodes so alternate that $\langle W_i, W_{i+1} \rangle$ is an edge of $\mathcal{G}$, for $i = 1, \ldots, k - 1$. A path is said to be SIMPLE if no place or node occurs twice in it, i.e., $W_i \neq W_j$ when $1 \leqslant i < j \leqslant k$.

If $k > 1$ and $W_1 = W_k$, the path $W_1, \ldots, W_k$ is a CYCLE; if, in addition, the vertices $W_1, \ldots, W_{k-1}$ are pairwise distinct, the cycle is SIMPLE. Plainly, the length of any cycle must be a positive even number.

Definition 5.4 (Pumping paths). *Relative to a greedy colored Π-process $\mathcal{H} = \left\langle \left(\Pi_\mu \right)_{\mu \leqslant \xi}, (\bullet), \mathcal{T}, \mathcal{R} \right\rangle$ over $\mathcal{G}$, a* PROSPECTIVE PUMPING PATH *$\mathcal{D}$ is a path in $\mathcal{G}$ of the form*

$$W_1, \ W_2, \ \ldots, \ W_n(= W_1), \ W_{n+1}, \ \ldots, \ W_m \tag{5.1}$$

(where $3 \leqslant n \leqslant m$) such that

- *$\mathcal{D}$ is devoid of red vertices;*

- *the vertex W_m is a place;*

- *$W_1, W_2, \ldots, W_n$ is a simple cycle, called the* PUMPING CYCLE OF *$\mathcal{D}$;*

- *the vertices $W_1, W_2, \ldots, W_{n-1}, W_{n+1}, \ldots, W_m$ are pairwise distinct, so that*

$$W_{n-1}, \ W_n, \ W_{n+1}, \ \ldots, \ W_m \tag{5.2}$$

is a simple path.

The TAIL *of* $\mathcal{D}$, *denoted by* tail$(\mathcal{D})$, *is the longest subpath of (5.2) starting with a place; thus, if W_{n-1} is a place, then the tail of $\mathcal{D}$ is just the path $W_{n-1}, W_n, W_{n+1}, \ldots, W_m$, otherwise it is the path $W_n, W_{n+1}, \ldots, W_m$.*

A prospective pumping path (5.1) is a PUMPING PATH FOR $\mathcal{H}$ *if its tail contains at least one place $q \in \Pi$ such that* leastInf(q) *is a limit ordinal. For a pumping path $\mathcal{D}$, we shall refer to the* last *place q in its tail such that* leastInf(q) *is a limit ordinal as* THE LIMIT PLACE OF THE PUMPING PATH $\mathcal{D}$. $\blacksquare$

We have basically two different kinds of cycles. One begins its tail with a place, the other with a node. Even if those phenomena are quite different in their behaviour, they yield to the same result, namely, the propagation of the infinite as the following example shows.

The situation radically changes when one has restrictions on the types of elements that can be distributed, as exemplified next.

Example 5.5. In the present example only unordered pairs can be distributed, and in fact there is a cycle $(\beta \to \{\alpha, \beta\})$ that generates infinite elements, which cannot be distributed along the tail, or, in other terms, there is a cycle that does not *propagate the infinite*.

Consider the following process

$$\begin{pmatrix} \alpha^{(\mu)} \\ \beta^{(\mu)} \\ \gamma^{(\mu)} \end{pmatrix}_{\mu \leq 4} = \left(\begin{array}{c|c|c|c|c} \emptyset & \emptyset^1 & \emptyset^1 & \emptyset^1 & \emptyset^1 \\ \emptyset & \emptyset & \emptyset^2 & \emptyset^2, \{\emptyset^1, \emptyset^2\} & \emptyset^2, \{\emptyset^1, \emptyset^2\} \\ \emptyset & \emptyset & \emptyset & \emptyset & \{\emptyset^1, \{\emptyset^1, \emptyset^2\}\} \end{array} \right),$$

with history

$$\begin{pmatrix} A_\mu \\ T_\mu \end{pmatrix}_{\mu < 4} = \left(\begin{array}{c|c|c|c} \emptyset & \{\alpha\} & \{\alpha, \beta\} & \{\alpha, \beta\} \\ \{\alpha\} & \{\beta\} & \{\beta\} & \{\gamma\} \end{array} \right),$$

and Π-board

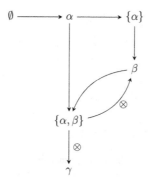

Let

$$\Phi \stackrel{\text{Def}}{:=} x \otimes y = z \wedge z \cap y \neq \emptyset.$$

The assignment

$$Mx := \alpha^{(4)}, \quad My := \beta^{(4)}, \quad Mz := \beta^{(4)} \cup \gamma^{(4)}$$

.

plainly satisfies φ.

Observe that the edges from node $\{\alpha, \beta\}$ are actually "Cartesian edges", namely, edges that can distribute just unordered pairs. Observe also that the following formula:

$$\Phi' \overset{\text{Def}}{:=} x \otimes y = z \wedge z \cap y \neq \emptyset \wedge \neg Finite(z)$$

cannot be satisfied starting from the present assignment, even if there is a cycle which supplies a region of Mz. Indeed, as noted above, this Π-board has a cycle that can be infinitely pumped, but cannot propagate the infinite along the tail even if there is no cardinal restriction. ∎

In what follows, we shall set up the ground for a theorem generalizing the following easy proposition (whose proof we delay until we can generalize the statement):

Lemma 5.6. *Relative to a greedy colored Π-process $\mathcal{H} = \langle (\Pi_\mu)_{\mu \leqslant \xi}, (\bullet), \mathcal{T}, \mathcal{R} \rangle$, consider a place s such that $|s^{(\bullet)}| \geqslant \aleph_0$. Then s lies on a pumping path for $\mathcal{H}$.*

In view of a proposition generalizing this one, we first prove the following two lemmas.

Lemma 5.7. *Relative to a greedy colored Π-process $\langle (\Pi_\mu)_{\mu \leqslant \xi}, (\bullet), \mathcal{T}, \mathcal{R} \rangle$, whose sequence of moves is $(A_\nu)_{\nu < \xi}$, consider a place s such that $\mathsf{leastInf}(s)$ is a successor ordinal. Then there exist a place g and a simple green path $\mathcal{E}$ in $\mathcal{G}$ from g to s such that*

(a) $\mathsf{leastInf}(g)$ *is a limit ordinal λ, and $\mathsf{leastInf}(q)$ is a successor ordinal, for every place $q \neq g$ in $\mathcal{E}$;*

(b) $\lambda \leqslant \mathsf{leastInf}(B) < \mathsf{leastInf}(q)$, *for every edge $\langle B, q \rangle$ lying on $\mathcal{E}$;*

(c) $\mathsf{leastInf}(q) \leqslant \mathsf{leastInf}(B)$, *for every edge $\langle q, B \rangle$ lying on $\mathcal{E}$;*

(d) *for every edge $\langle B, q \rangle$ lying on $\mathcal{E}$ and for every node H other than B, it holds that*

$$\left| q^{(\mathsf{leastInf}(q))} \cap \mathcal{P}^* \left(B^{(\mathsf{leastInf}(q)-1)} \right) \right| \geqslant \aleph_0 \qquad and$$
$$\left| q^{(\mathsf{leastInf}(q))} \cap \mathcal{P}^* \left(H^{(\mathsf{leastInf}(q)-1)} \right) \right| < \aleph_0.$$

Proof. Arguing by contradiction, let us assume that $\bar{\nu}$ be the least ordinal such that an s with $\mathsf{leastInf}(s) = \bar{\nu} + 1$ exists violating the thesis. (Clearly s must be green.) Then $\left| \mathcal{P}^* \left(A_{\bar{\nu}}^{(\bar{\nu})} \right) \right| \geqslant \aleph_0$, where $A_{\bar{\nu}}$ is the $\bar{\nu}$-th move, and hence $A_{\bar{\nu}}$ is green and there is a green place $s' \in A_{\bar{\nu}}$ such that

$$0 < \mathsf{leastInf}(s') \leqslant \mathsf{leastInf}(A_{\bar{\nu}}) \leqslant \bar{\nu} < \bar{\nu} + 1 = \mathsf{leastInf}(s). \tag{5.3}$$

It can be easily checked that the edge $\langle A_{\bar{\nu}}, s \rangle$ satisfies condition (d). Thus, if $\mathsf{leastInf}(s')$ were a limit ordinal, our initial assumption would be violated, because we could take $\mathcal{E} \equiv s', A_{\bar{\nu}}, s$ (conditions (a), (b), and (c) would readily follow from (5.3)). Hence $\mathsf{leastInf}(s') = \gamma + 1$, and therefore, by the assumed minimality of $\bar{\nu}$, a green path $\mathcal{E}'$ leading to s' and satisfying the conditions of the lemma can be found. But then, by prolonging $\mathcal{E}'$ first with $A_{\bar{\nu}}$ and next with s, we would find a simple path $\mathcal{E}$ whose existence conflicts with our initial assumption. □

Lemma 5.8. *Consider a greedy colored* Π*-process* $\left\langle \left(\Pi_\mu\right)_{\mu \leqslant \xi}, (\bullet), \mathcal{T}, \mathcal{R} \right\rangle$ *over a* Π*-board* $\mathcal{G} = \langle \mathcal{T}, \mathcal{R} \rangle$, *whose sequence of moves is* $\left(A_\nu\right)_{\nu < \xi}$, *and let* λ *be a limit ordinal. Let, moreover,* $\mathcal{G}^\lambda$ *be the subgraph of* $\mathcal{G}$ *whose vertices are*

- *all places* $q \in \Pi$ *such that the set*
$$I_q^\lambda := \{\nu < \lambda \mid q^{(\nu+1)} \neq q^{(\nu)}\}$$
is cofinal in λ *(i.e.,* $\bigcup I_q^\lambda = \lambda$), *and*

- *all nodes* $B \subseteq \Pi$ *such that the set*
$$I_B^\lambda := \{\nu < \lambda \mid A_\nu = B\}$$
is cofinal in λ,

and whose edges are

- *all pairs* $\langle B, q \rangle$ *such that*
$$I_{B,q}^\lambda := I_B^\lambda \cap I_q^\lambda = \{\nu < \lambda \mid A_\nu = B \wedge q^{(\nu+1)} \neq q^{(\nu)}\}$$
is cofinal in λ,[2] *and*

- *all pairs* $\langle q, B \rangle$ *such that* $q \in B$, *where the place* q *and node* B *are vertices of* $\mathcal{G}^\lambda$.

Then every vertex W *of* $\mathcal{G}^\lambda$, *if any exists, has at least one predecessor in* $\mathcal{G}^\lambda$ *and satisfies* $0 < \mathsf{leastInf}(W) \leqslant \lambda$ *(and therefore it is green).*

Proof. If W is a *place* p of $\mathcal{G}^\lambda$ then, since

$$I_p^\lambda = \bigcup \{I_{B,p}^\lambda : B \subseteq \Pi\},$$

from the assumption $\bigcup I_p^\lambda = \lambda$ and the finiteness of Π it follows that $\bigcup I_{B,p}^\lambda = \lambda$ for some $B \subseteq \Pi$. Hence B is an immediate predecessor of p in $\mathcal{G}^\lambda$. Also, since $|p^{(\lambda)}| \geqslant \aleph_0$, we have $0 < \mathsf{leastInf}(p) \leqslant \lambda$.

On the other hand, if W is a *node* B of $\mathcal{G}_\lambda$ then, from the assumption $\bigcup I_B^\lambda = \lambda$ and the greediness of the process, it follows that $\bigcup \bigcup \{I_q^\lambda : q \in B\} = \lambda$. Therefore $\bigcup I_q^\lambda = \lambda$ for some $q \in B$, in view of the finiteness of B, and hence q is an immediate predecessor of B in $\mathcal{G}^\lambda$. In addition, we have $|\mathcal{P}^*(B^{(\lambda)})| \geqslant \aleph_0$, and therefore $0 < \mathsf{leastInf}(B) \leqslant \lambda$. $\square$

The preceding two lemmas imply the following basic result, which will be generalized in Section 5.4.1.

Theorem 5.9 (Pumping Path Theorem). *Relative to a greedy colored* Π*-process* $\mathcal{H} = \left\langle \left(\Pi_\mu\right)_{\mu \leqslant \xi}, (\bullet), \mathcal{T}, \mathcal{R} \right\rangle$ *over a* Π*-board* $\mathcal{G} = \langle \mathcal{T}, \mathcal{R} \rangle$, *whose sequence of moves is* $\left(A_\nu\right)_{\nu < \xi}$, *consider a place* s *such that* $|s^{(\bullet)}| \geqslant \aleph_0$. *Then* s *is the last node of a pumping path* $\mathcal{D}$ *for* $\mathcal{H}$ *whose tail contains a place* g, *with* $\mathsf{leastInf}(g)$ *a limit ordinal* λ, *and such that the following conditions hold:*

1. $0 < \mathsf{leastInf}(q)$, $\mathsf{leastInf}(B) \leqslant \lambda = \bigcup I_{B,q}^\lambda$ *holds, where* $I_{B,q}^\lambda = \{\nu < \lambda \mid A_\nu = B \wedge q^{(\nu+1)} \neq q^{(\nu)}\}$, *for every edge* $\langle B, q \rangle$ *preceding* g *on* $\mathcal{D}$;

---

[2]Observe that if, for a node B and a place q of $\mathcal{G}$, $I_{B,q}^\lambda$ is cofinal in λ, then from $I_{B,q}^\lambda \subseteq I_B^\lambda \subseteq \lambda$ and $I_{B,q}^\lambda \subseteq I_q^\lambda \subseteq \lambda$ it follows that I_B^λ and I_q^λ are both cofinal in λ, so that B and q are vertices of $\mathcal{G}^\lambda$.

2. $\lambda \leqslant \mathsf{leastInf}(B) < \mathsf{leastInf}(q)$, *for every edge* $\langle B, q \rangle$ *following* g *on* $\mathcal{D}$;

3. $\mathsf{leastInf}(q) \leqslant \mathsf{leastInf}(B)$, *for every edge* $\langle q, B \rangle$ *following* g *on* $\mathcal{D}$;

4. *for every edge* $\langle D, p \rangle$ *following* g *on* $\mathcal{D}$ *and for every node* H *other than* D, *it holds that*

$$\left| p^{(\mathsf{leastInf}(p))} \cap \mathcal{P}^*\big(D^{(\mathsf{leastInf}(p)-1)}\big) \right| \;\geqslant\; \aleph_0 \qquad and$$
$$\left| p^{(\mathsf{leastInf}(p))} \cap \mathcal{P}^*\big(H^{(\mathsf{leastInf}(p)-1)}\big) \right| \;<\; \aleph_0 \,.$$

Proof. If $\mathsf{leastInf}(s)$ is a limit ordinal λ, we construct $\mathcal{D}$ by exploiting Lemma 5.8, where $\mathcal{G}^\lambda$ turns out to be non-void since it contains the place s. Thanks to the fact that every vertex of $\mathcal{G}^\lambda$ has a predecessor in $\mathcal{G}^\lambda$, starting with s and proceeding in a backward fashion in $\mathcal{G}^\lambda$, eventually we shall hit the same node of $\mathcal{G}^\lambda$ twice, since $\mathcal{G}^\lambda$ is finite. Plainly, the sequence of vertices so encountered, but in reverse order, is a pumping path to s, having as limit node s itself. By putting $g := s$, conditions 2, 3, and 4 are then vacuously true, and condition 1 is an immediate consequence of Lemma 5.8.

If $\mathsf{leastInf}(s)$ is a successor ordinal, then by Lemma 5.7 there exists a simple green path $\mathcal{E}$ from a place g to s such that

(i) $\mathsf{leastInf}(g)$ is a limit ordinal λ, and $\mathsf{leastInf}(q)$ is a successor ordinal, for every place $q \neq g$ in $\mathcal{E}$;

(ii) $\lambda \leqslant \mathsf{leastInf}(B) < \mathsf{leastInf}(q)$, for every edge $\langle B, q \rangle$ lying on $\mathcal{E}$;

(iii) $\mathsf{leastInf}(q) \leqslant \mathsf{leastInf}(B)$, for every edge $\langle q, B \rangle$ lying on $\mathcal{E}$;

(iv) for every edge $\langle B, q \rangle$ lying on $\mathcal{E}$ and for every node H other than B, it holds that

$$\left| q^{(\mathsf{leastInf}(q))} \cap \mathcal{P}^*\big(B^{(\mathsf{leastInf}(q)-1)}\big) \right| \;\geqslant\; \aleph_0 \qquad and$$
$$\left| q^{(\mathsf{leastInf}(q))} \cap \mathcal{P}^*\big(H^{(\mathsf{leastInf}(q)-1)}\big) \right| \;<\; \aleph_0 \,.$$

To construct the pumping path $\mathcal{D}$ of the theorem, we argue as follows. Starting from g, we construct a path $\mathcal{E}'$ to g in $\mathcal{G}^\lambda$ by proceeding backwards in $\mathcal{G}^\lambda$ until either we hit a vertex W on $\mathcal{E}$ distinct from g or we hit a same vertex twice. In the former case, the vertex W can only be the successor of g in $\mathcal{E}$. Indeed, from Lemma 5.8, we have $0 < \mathsf{leastInf}(W) \leqslant \lambda$, whereas, by (ii) and (iii) above, we have $\mathsf{leastInf}(W') > \lambda$, for every vertex W' on $\mathcal{E}$ distinct from g and from the immediate successor of g in $\mathcal{E}$.

Let $\mathcal{D}$ be the path obtained by gluing $\mathcal{E}$ at the end of the path $\mathcal{E}'$ found in $\mathcal{G}^\lambda$.

In view of (i)–(iv) above and Lemma 5.8, it is an easy matter to verify that $\mathcal{D}$ is a pumping path in $\mathcal{G}$ with limit node g, satisfying the conditions of the theorem. □

5.3.2 Local Trashes

Consider a colored Π-board $\mathcal{G} = (\mathcal{T}, \mathcal{R})$ complying with a $\mathcal{P}$-partition Σ with special block $\overline{\sigma}$ via $q \mapsto q^{(\bullet)}$. We recall that a $\mathcal{P}$-node is any node of $\mathcal{G}$ not containing the special place $\overline{q} \in \Pi$ such that $\overline{q}^{(\bullet)} = \overline{\sigma}$. Thus, any $\mathcal{P}$-node needs to be totally drained before the end of the formative process, otherwise the property to be a $\mathcal{P}$-partition turns out to be disrupted. In what follows, we refine into a useful proposition the following easy remark: the block at a place s belonging to a $\mathcal{P}$-node A cannot become infinite during a colored process unless A has a green place among its targets.

To see that such kind of target must exist, assume that A is a $\mathcal{P}$-node containing an element $s \in A$ such that $\left|s^{(\bullet)}\right| \geqslant \aleph_0$. Consequently, $\left|\mathcal{P}^*(A^{(\bullet)})\right| > \aleph_0$ and $\mathcal{P}^*(A^{(\bullet)}) \subseteq \bigcup \Pi^{(\bullet)}$ (by Property (II) of $\mathcal{P}$-partition), and hence there must be a place $g \in \Pi$ such that $\left|\mathcal{P}^*(A^{(\bullet)}) \cap g^{(\bullet)}\right| > \aleph_0$, because $\left|\Pi^{(\bullet)}\right| = |\Pi| < \aleph_0$, which obviously implies that $g \in \mathcal{T}(A) \setminus \mathcal{R}$.[3]

We first give the definition of *local trashes*, namely, places enjoying suitable properties which allow one to "drain off" the "residual" elements of a pumping process.

Definition 5.10 (Local trashes). *Relative to a greedy colored* Π-*process* $\left\langle (\Pi_\mu)_{\mu \leqslant \xi}, (\bullet), \mathcal{T}, \mathcal{R} \right\rangle$, *a place g is a* LOCAL TRASH *for a node $A \subseteq \Pi$ if*

- $g \in \mathcal{T}(A) \setminus \mathcal{R}$, *i.e.*, g *is a green target of A;*

- *every node $B \subseteq \Pi$ such that $g \in B$ has* $\mathsf{GE}(B) > \mathsf{GE}(A)$. ∎

A basic fact, which we prove in the following technical lemma, is that every $\mathcal{P}$-node A such that $\bigcup A^{(\bullet)}$ is infinite has an infinite local trash.

A simple way to establish whether there exists or not the above mentioned partition is just to check whether the fixed cycle possess a closure. A possible procedure to detect this could be LocalTrash in Section 5.7. If the procedure answers 'yes' and returns the set P then, as a by-product, $\mathcal{R} = \Pi \setminus \mathsf{P}$. If a fixed language has the property that all cycles possess automatically a closure for that language, the notion of dynamical cycle coincides with that of the statical one.

Lemma 5.11. *Relative to a greedy colored Π-process* $\left\langle (\Pi_\mu)_{\mu \leqslant \xi}, (\bullet), \mathcal{T}, \mathcal{R} \right\rangle$, *consider a $\mathcal{P}$-node A for which $\bigcup A^{(\bullet)}$ is infinite. Then A has a local trash p such that $p^{(\bullet)}$ is infinite.*

Proof. By the greediness of the process, since A is a $\mathcal{P}$-node, we have $\mathcal{P}^*(A^{(\alpha)}) = \mathcal{P}^*(A^{(\bullet)}) \subseteq \bigcup \Pi^{(\bullet)}$; hence $\mathcal{P}^*(A^{(\alpha)}) \setminus \bigcup \Pi^{(\alpha)} \subseteq \bigcup \Pi^{(\alpha+1)}$, again by the greediness of the process. Moreover, since $\mathcal{P}^*(A^{(\alpha)}) \setminus \bigcup \Pi^{(\alpha)}$ is an infinite set by Lemma 3.20, there must be a place $p \in \mathcal{T}(A)$ for which the set $p^{(\alpha+1)} \cap (\mathcal{P}^*(A^{(\alpha)}) \setminus \bigcup \Pi^{(\alpha)})$ is infinite too. Thus, $p^{(\bullet)}$ is infinite, so that $p \notin \mathcal{R}$. In addition, since $p^{(\alpha+1)} \setminus p^{(\alpha)} \neq \emptyset$, it readily follows that $\mathsf{GE}(B) > \alpha = \mathsf{GE}(A)$ must hold for every node $B \subseteq \Pi$ such that $p \in B$, which proves that p is a local trash for A. □

If we put $\mathsf{P} := \{q \in \Pi \mid q^{(\bullet)} \text{ is infinite}\}$, then Lemma 5.11 can be rephrased in the following way:

$(\star)$ every $\mathcal{P}$-node A with a non-null intersection with the set P has a local trash inside P.

In general, given a colored Π-board $\langle \mathcal{T}, \mathcal{R} \rangle$, a set of places $\mathsf{P} \subseteq \Pi \setminus \mathcal{R}$ which enjoys property $(\star)$ is said to be $\mathcal{P}$-CLOSED|TEXTBF (thus, Lemma 5.11 simply says that the set of 'infinite' places $\{q \in \Pi \mid q^{(\bullet)}\}$ is $\mathcal{P}$-closed).

---

[3] As already remarked, MLSSP reflects over the hereditarily finite sets. Thus, in a case like this, whenever no literal of type $y = \neg Finite(x)$ is present in a conjunction Φ of MLSSPF, it will always be possible to extract, out of a given formative process for a model for Φ, a shadow process of finite length related to a hereditarily finite model for Φ.

5.4   Shadow Model and Potentially Infinite Venn Regions

Here we introduce some kind of "dynamical" pumping path, namely, the *pumping chain*. This object is, in a sense, a pumping path observed as it is created. Indeed, the pumping path shows its capability to create infinite elements only in a specific moment of the formative process. The synchronicity of the original process and the shadow process allows us to prove that these moments are exactly the same, up to the bijection between the original process and the shadow process. For this reason, we do not need a new process in order to preserve the pumping chains of the original process. In fact, the shadow process created in the previous chapter already preserves the pumping chains of the original process. The above mentioned synchronicity property guarantees that the new process has the same pumping chains as the original one, triggered exactly in the same steps, as can be seen in the shadow process. Moreover, this result implies the capability to create a model for a MLSSP-formula Φ which reveals, at the same time, which variable $v \in \mathrm{Vars}(\Phi)$ can be made infinite and which can not. For this reason, the properties we are going to prove will enlarge the information available through the model depicted in the previous chapter. This information lies inside the same model which allowed us to say that Φ is satisfiable. In a sense MLSSPF is a pure virtual extension of MLSSP.

5.4.1   Pumping Chains

We can introduce now a very important combinatorial structure, namely, *pumping chains*, and then prove that the shadow $\widehat{\Pi}$-process $\left\langle (\widehat{\Pi}_i)_{i \leqslant \ell}, [\bullet], \widehat{\mathcal{T}}, \widehat{\mathcal{R}} \right\rangle$ of a formative Π-process $\left\langle (\Pi_\mu)_{\mu \leqslant \xi}, (\bullet), \mathcal{T}, \mathcal{R} \right\rangle$ possesses a pumping chain to each place $\widehat{p} \in \widehat{\Pi}$ such that $p^{(\bullet)}$ is infinite (*Pumping Chain Theorem*). As will be shown then in Section 5.5.2, pumping chains can be used to generate infinite elements and propagate them to the blocks associated with their places, while preserving L-simulation.

Given a path $\mathcal{D}$ in a Π-board $\mathcal{G}$, we denote with $(\mathcal{D})_{places}$ and $(\mathcal{D})_{nodes}$ the collections of the places and of the nodes occurring in $\mathcal{D}$, respectively. Moreover, given a node B in $\mathcal{G}$, we designate by $\mathcal{N}(B)$ the collection of all the nodes which have non-null intersection with B.

Definition 5.12 (Pumping chains). *Let* $\mathcal{H} = \left\langle (\Pi_\mu)_{\mu \leqslant \xi}, (\bullet), \mathcal{T}, \mathcal{R} \right\rangle$ *be a greedy colored process over a colored Π-board* $\mathcal{G} = \langle \mathcal{T}, \mathcal{R} \rangle$.

A PUMPING CHAIN *in* $\mathcal{H}$ *to a place* $q \in \Pi$ *is a quadruple* $\mathcal{PC} = \langle \mathcal{D}, \eta, q_0, \mathsf{P} \rangle$, *where*

- $\mathcal{D} = \langle \mathcal{D}_0, \mathcal{D}_1, \ldots, \mathcal{D}_N \rangle$ *is a partition into contiguous paths, each ending with a place, of a prospective pumping path* $\mathcal{D}$ *to* q *in* $\mathcal{G}$, *where*

$$\mathcal{D} \equiv W_1, \ldots, W_{n-1}, W_n(= W_1), W_{n+1}, \ldots, W_r(= q)$$

  (with pumping cycle $\mathcal{C} \equiv W_1, \ldots, W_{n-1}, W_n$),

- $\eta = \langle \eta_0, \eta_1, \ldots, \eta_N \rangle$ *is a stricly increasing sequence of non-null ordinals not greater than* ξ, *i.e.,* $0 < \eta_0 < \eta_1 < \eta_2 < \ldots < \eta_N < \xi$, *with* $\xi - \eta_0$ *a finite ordinal (thus,* $\eta_1, \eta_2, \ldots, \eta_N, \xi$ *are successor ordinals),*

- $q_0 \in (\mathcal{C})_{places}$,

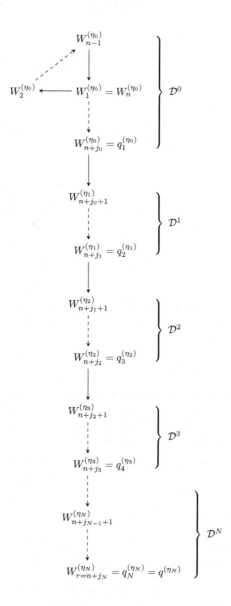

Figure 5.1: A pumping chain.

- P *is a* $\mathcal{P}$*-closed set of places such that every* $\mathcal{P}$*-node* B *that intersects* P *has* $\mathsf{GE}(B) \geq \eta_0$ *(cf. definition at the end of Section 5.3.2),*[4]

such that the following conditions are satisfied:

(i) $q_0^{(\eta_0)} \setminus q_0^{(\eta_0-1)} \neq \emptyset$,

(ii) *the path* $\mathcal{D}_0$ *contains the initial subpath* $\langle W_1, \ldots, W_{n-1} \rangle$ *of* $\mathcal{D}$,

(iii) $\mathsf{GE}\big(\mathcal{N}\big((\mathcal{D}_i)_{pl}\big)\big) \geqslant \eta_i$, *for* $i = 0, 1, \ldots, N$,

(iv) $\mathcal{P}^*\big(B^{(\eta_i)}\big) \neq \emptyset$ *(i.e.,* $\emptyset \notin B^{(\eta_i)}$*), for* $B \in (\mathcal{D}_i)_{nd}$ *and* $i = 0, 1, \ldots, N$,

(v) $(\mathcal{D})_{places} \subseteq$ P.

The sequence η *and the ordinal* η_0 *will be called, respectively, the* time sequence *and the* starting time *of the pumping chain* $\mathcal{PC}$. *We also define the* LENGTH *of the pumping chain* $\mathcal{PC}$ *as the number of nodes in the tail of its pumping path; in symbols, we put* $\mathsf{length}(\mathcal{PC}) := |(\mathsf{tail}(\mathcal{D}))_{nodes}|$. ∎

We are now ready to show that our shadow process preserves pumping chains.

Theorem 5.13 (Pumping Chain Theorem). *Let* $\mathcal{H} = \big\langle (\Pi_\mu)_{\mu \leqslant \xi}, (\bullet), \mathcal{T}, \mathcal{R} \big\rangle$ *be a greedy colored process, whose sequence of moves is* $(A_\nu)_{\nu < \xi}$, *and let* $\widehat{\mathcal{H}} = \big\langle (\widehat{\Pi}_i)_{i \leqslant \ell}, [\bullet], \widehat{\mathcal{T}}, \widehat{\mathcal{R}} \big\rangle$ *be a shadow* $\widehat{\Pi}$*-process of* $\mathcal{H}$ *with respect to the salient sequence* $(\mu_i)_{i \leqslant \ell}$ *of steps of* $(\Pi_\mu)_{\mu \leqslant \xi}$ *(relative to some threshold* ϱ*).*

Then, for every $p \in \Pi$ *such that* $p^{(\bullet)}$ *is infinite,* $\widehat{\mathcal{H}}$ *owns a pumping chain to the place* $\widehat{p}$.

The details of the proof can be found in [CU14], nonetheless we briefly outline the main ideas. Let $p \in \Pi$ such that $p^{(\bullet)}$ is infinite. By Theorem 5.9, the process contains a pumping path $\mathcal{D}$ to p, with pumping cycle $\mathcal{C}$ and limit place g, such that, if we denote by λ the limit ordinal $\mathsf{leastInf}(g)$, the following properties hold:

(a) $0 < \mathsf{leastInf}(q)$, $\mathsf{leastInf}(B) \leqslant \lambda = \bigcup I_{B,q}^\lambda$, for every edge $\langle B, q \rangle$ preceding g on $\mathcal{D}$, where we recall that $I_{B,q}^\lambda = \{\nu < \lambda \mid A_\nu = B \wedge q^{(\nu+1)} \neq q^{(\nu)}\}$;

(b) $\lambda \leqslant \mathsf{leastInf}(B) < \mathsf{leastInf}(q)$, for every edge $\langle B, q \rangle$ following g on $\mathcal{D}$;

(c) $\mathsf{leastInf}(q) \leqslant \mathsf{leastInf}(B)$, for every edge $\langle q, B \rangle$ following g on $\mathcal{D}$.

The proof run throughout two important steps:

- showing that the known pumping path is actually a pumping chain;

- showing that the discovered pumping chain in the original process is actually a pumping chain for the shadow process, as well (see Figure 5.12).

---

[4]In view of Lemma 5.11 and condition (v) (which says that $(\mathcal{D})_{places} \subseteq$ P), P is a superset of the collection of the places that might need to become infinite, if all places in $\mathcal{D}$ were made infinite (say, by a pumping process of some sort).

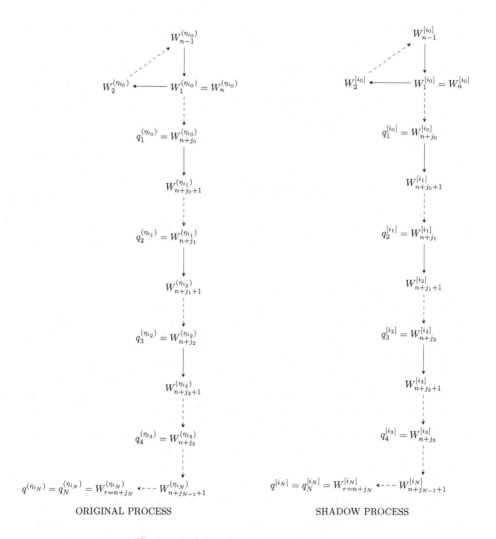

ORIGINAL PROCESS   SHADOW PROCESS

Figure 5.2: Original and shadow processes.

In order to handle the first step, we provide a partition $\overline{\mathcal{D}}$ of $\mathcal{D}$ into contiguous paths (each ending with a place) and an integer tuple $\overline{\eta}$ of the same size, a place q_0 in the pumping cycle of $\mathcal{D}$, and a $\mathcal{P}$-closed set of places P, satisfying conditions (i)–(v) of Definition 5.12.

By pure combinatorial arguments, we can find the sequences of ordinals that show the real capability of the cycle to distribute an infinite amount of elements. This is not a big surprise, since quite a few standard arguments drive to the requested properties However, what we are really looking for is a pumping chain in the shadow process. Therefore the pumping chain in the original process is simply a key to discover the desired object, which is the goal of the second step. Assuming we already have the collection of blocks, places and ordinals which satisfy the pumping chain conditions with respect to the pumping chain $\mathcal{PC} = \langle \mathcal{D}, \eta, q_0, \mathsf{P} \rangle$, we have to rephrase the pumping chain conditions in terms of the shadow process. In other terms every ordinal inside the definition of pumping chain, say, μ, must be an ordinal μ_i, selected by the shadow process (or at least μ has to determine univocally a μ_i). If this is the case, it turns to be rather natural to rephrase the above conditions in terms of shadow process, since it is sufficient to substitute all μ_i with i (see Figure 5.12). Obviously nothing guarantees in general that if, for instance, $\eta_j = \mu_i$ and $\mathcal{P}^*\big(B^{(\eta_j)}\big) \neq \emptyset$, then automatically we have $\mathcal{P}^*\big(B^{(i)}\big) \neq \emptyset$. Why should it be true? Exactly because of synchronicity properties of the shadow process. More precisely, the (promised) synchronization map $i \mapsto \mu_i$ is order-preserving since $\eta = \langle \eta_0 = \mu_{i_0}, \eta_1 = \mu_{i_1}, \ldots, \eta_N = \mu_{i_N} \rangle$ is a strictly increasing sequence of non-null ordinals. Therefore, $i = \langle i_0, i_1 \ldots i_N \rangle$ is a strictly increasing sequence of non-null integers too. Furthermore, the synchronicity of the shadow process guarantees properties (i)-(v).

5.5   A Decision Procedure for MLSSPF

We are now ready to refine the paradigmatic satisfiability test FirstSatTestMLSSPF, outlined in Section 5.1, into the fully-fledged decision procedure SatifisfiabilityTestMLSSPF for MLSSPF, reported in Table 5.1.

By exploiting the results presented in the first part of the chapter, we will soon check that the procedure SatifisfiabilityTestMLSSPF outputs the correct answer, whenever it is run with a satisfiable MLSSPF-conjunction (i.e., the procedure is *sound*). The converse result, namely that the procedure SatifisfiabilityTestMLSSPF will also output the correct answer whenever it is run with an unsatisfiable MLSSPF-conjunction (*completeness*), will be addressed in the subsequent section. Since the procedure SatifisfiabilityTestMLSSPF always terminates, as is immediate to check, it will therefore follow that the satisfiability problem for MLSSPF is decidable.

5.5.1   Soundness of the Procedure SatifisfiabilityTestMLSSPF

Let Φ be a satisfiable normalized MLSSPF-conjunction, and M a model for it. Let V_Φ, m, Fin_Φ, notFin_Φ, Enum_Φ, and ℓ_Φ be as in the procedure SatifisfiabilityTestMLSSPF. Let Σ_M be the Venn partition for Φ and $\mathrm{Vars}(\Phi)$, and $\mathcal{T}$ the target function of the associated Σ_M-board $\mathcal{G}$, with set of places Π. Plainly, $|\Pi| = |\Sigma_M| \leq 2^m$. Let $\mathcal{H} = \langle (\Pi_\mu)_{\mu \leqslant \xi}, (\bullet), \mathcal{T} \rangle$ be a greedy Π-process which ends with $\Pi_\xi = \Sigma_M$, and $\mathcal{I} : V_\Phi \to \mathcal{P}(\Pi)$ a map such that $\mathcal{I}(x) = \{q \in \Pi \mid q^{(\bullet)} \subseteq M\,x\}$, for $x \in V_\Phi$. Let $\varrho = \max(m, 3) + m + 1$, so that $\varrho > m - 1$ and $2^{\varrho - 1} > \varrho \cdot 2^m \geqslant \varrho \cdot |\Pi|$. By applying the procedure imitate of [COU02] on $\mathcal{H}$, it turns

```
procedure SatifisfiabilityTestMLSSPF(Φ);
        V_Φ := set of variables occurring in Φ;
        m := |V_Φ|;
        Fin_Φ := {x ∈ V_Φ | Finite(x) occurs in Φ};
        notFin_Φ := {x ∈ V_Φ | ¬Finite(x) occurs in Φ};
        Enum_Φ := {x ∈ V_Φ | a literal of the form x = {y_1,..., y_h} occurs in Φ};
        ℓ_Φ := ⌈25/24 m + 5⌉ · 2^{2^m+m} + 3;
        for   - each set of places Π of cardinality at most 2^m,
              - each Π-board 𝒢 (with target function 𝒯 ∈ 𝒫(Π)^{𝒫(Π)}),
              - each formative process ζ = ⟨(Π_μ)_{μ≤ℓ}, (•), 𝒯⟩ over 𝒢 such that ℓ < ℓ_Φ, and
              - each map ℑ : V_Φ → 𝒫(Π)    do
        M_{ζ,ℑ} := set assignment over V_Φ such that M_{ζ,ℑ} x = ⋃(ℑ(x))^{(•)}, for x ∈ V_Φ;
        if M_{ζ,ℑ} does not satisfy Φ^- then continue; end if;
        ℛ := ⋃{ℑ(x) | x ∈ Fin_Φ ∪ Enum_Φ};
        if for each x ∈ notFin_Φ there exists a pumping chain in the colored Π-process
                ⟨(Π_μ)_{μ≤ℓ}, (•), 𝒯, ℛ⟩ through some place in ℑ(x) then
            return "Φ is satisfiable";
        else continue;
        end if;
    end for;
    return "Φ is unsatisfiable";
end procedure;
```

Table 5.1: A decision procedure for the satisfiability problem for MLSSPF.

out that there exists a (weak) Π-process $\widehat{\mathcal{H}} = \langle(\widehat{\Pi}_i)_{i\leq\ell}, [\bullet], \widehat{\mathcal{T}}\rangle$, relative to the threshold ϱ, such that

- $\ell < \max\left(\lceil\frac{25}{24} \cdot \lceil\log_2|\Sigma_M|\rceil + 5\rceil, m\right) \cdot |\Sigma_M| \cdot 2^{|\Sigma_M|} + 3$
  $\leq \lceil\frac{25}{24}m + 5\rceil \cdot 2^{2^m+m} + 3 = \ell_\Phi$, and

- $\widehat{\Pi}_\ell$ $(m-1)$-simulates Σ_M.

Since no literal of the form $x = \{y_1,\ldots,y_h\}$ in Φ, with $y_1,\ldots,y_h$ distinct variables, can involve on its right-hand side more than $m-1$ variables,[5] it follows that the set assignment $M_{\widehat{\mathcal{H}},\widehat{\mathcal{J}}}$ over V_Φ, where $M_{\widehat{\mathcal{H}},\widehat{\mathcal{J}}} x = \bigcup(\widehat{\mathcal{J}}(x))^{[\bullet]}$ (with $\widehat{\mathcal{J}} \in \mathcal{P}(\widehat{\Pi})^{V_\Phi}$ the map such that $\widehat{\mathcal{J}}(x) = \{\widehat{q} \mid q^{(\bullet)} \subseteq M x\}$), is a model for Φ^-. Finally, let $\widehat{\mathcal{R}} := \bigcup\{\widehat{\mathcal{J}}(x) \mid x \in \text{Fin}_\Phi \cup \text{Enum}_\Phi\}$. It is immediate to check that the Π-process $\widehat{\mathcal{H}}$ complies with the colored Π-board $\langle\widehat{\mathcal{T}}, \widehat{\mathcal{R}}\rangle$. Thus, by Theorem 5.13, for each $p \in \Pi$ such that $p^{(\bullet)}$ is infinite, $\widehat{\mathcal{H}}$ owns a pumping chain to $\widehat{p}$. But since for each $x \in \text{notFin}_\Phi$ there exists $p \in \mathcal{J}(x)$ such that $p^{(\bullet)}$ is infinite, the condition of the **if-then** statement of procedure SatifisfiabilityTestMLSSPF at lines 13–16 is satisfied, and therefore the conjunction Φ is declared 'satisfiable' by procedure SatifisfiabilityTestMLSSPF, which is what we intended to prove.

In the following section we shall argue that whenever the procedure SatifisfiabilityTestMLSSPF declares that its input formula is satisfiable, this is indeed the case, or, equivalently, that the procedure SatifisfiabilityTestMLSSPF will also output the correct answer whenever it is run with an unsatisfiable MLSSPF-conjunction.

---

[5]Otherwise, x would occur among the variables $y_1,\ldots,y_h$ and the literal $x = \{y_1,\ldots,y_h\}$ would be unsatisfiable by itself.

5.5.2 Completeness of the Procedure SatifisfiabilityTestMLSSPF

We provide now the second half of the correctness proof of the decision procedure for MLSSPF in Table 5.1, by proving its completeness. This amounts to showing that whenever the procedure SatifisfiabilityTestMLSSPF declares that its input MLSSPF-conjunction is satisfiable, this is indeed the case.

Thus, let Φ be a normalized MLSSPF-conjunction and let V_Φ, m, Fin_Φ, notFin_Φ, Enum_Φ, and ℓ_Φ be as in the procedure SatifisfiabilityTestMLSSPF. As before, let Φ^- be the conjunction obtained by dropping from Φ all the literals of any of the forms $Finite(x)$ and $\neg Finite(x)$.

When we proceed in an exhaustive screening of all possible assignments of all possible transitive partitions with an associated formative partition of a fixed rank, at first we detect if it is a model for Φ^-. Even if this would be the case, we are still left to create a new model using the structural properties that we have found in order to make infinite the assignment to the variables which are supposed to be so. In other terms, in order to prove the completeness, first we have to check whether an assignment gives rise to a model (for Φ^-), then we have to create a new model (for Φ).

We denote by Φ_{Fin} and by $\Phi_{\neg\mathsf{Fin}}$ the conjunctions of the literals of type $\mathsf{Finite}(x)$ and of type $\neg\mathsf{Finite}(x)$ in Φ, respectively. Let also Π be a set of places of cardinality at most 2^m, $\mathcal{G}$ a Π-board with target function $\mathcal{T} \in \mathcal{P}(\Pi)^{\mathcal{P}(\Pi)}$, $\zeta = \langle (\Pi_\mu)_{\mu \leqslant \ell}, (\bullet), \mathcal{T} \rangle$ a formative process over $\mathcal{G}$ with $\ell < \ell_\Phi$, $\mathcal{I} : V_\Phi \to \mathcal{P}(\Pi)$ a map, $M_{\zeta,\mathcal{I}}$ the set assignment over V_Φ such that $M_{\zeta,\mathcal{I}} x = \bigcup (\mathcal{I}(x))^{(\bullet)}$ and $\mathcal{R} := \bigcup \{\mathcal{I}(x) \mid x \in \mathsf{Fin}_\Phi \cup \mathsf{Enum}_\Phi\}$, and assume that

(a) the set assignment $M_{\zeta,\mathcal{I}}$ satisfies Φ^-, and

(b) for each $x \in \mathsf{notFin}_\Phi$, there exists a pumping chain through some place in $\mathcal{I}(x)$ in the colored Π-process $\langle (\Pi_\mu)_{\mu \leqslant \ell}, (\bullet), \mathcal{T}, \mathcal{R} \rangle$.

Observe that these are just the conditions under which the procedure SatifisfiabilityTest-MLSSPF would declare the conjunction Φ to be satisfiable.

Let PCS be a sequence of pumping chains in ζ which *cover* all the variables in notFin_Φ, in the sense that, for each $x \in \mathsf{notFin}_\Phi$, it contains a pumping chain through some place $q \in \Pi$ such that $q^{(\bullet)} \subseteq M_{\zeta,\mathcal{I}} x$. (The existence of the sequence PCS is ensured from the previous condition (b).)

We shall show that, under the above conditions, $M_{\zeta,\mathcal{I}}$ can be suitably modified in such a way that the resulting assignment M satisfies all the conjuncts of type $Finite(x)$ and $\neg Finite(x)$ occurring in Φ, in addition to those in Φ^-. The idea underlying this pumping procedure is rather intuitive:

1. Pick a variable x in notFin_Φ. By condition (b) there exists a pumping chain through some place σ in $\mathcal{I}(x)$ in the colored Π-process $\langle (\Pi_\mu)_{\mu \leqslant \ell}, (\bullet), \mathcal{T}, \mathcal{R} \rangle$.

2. "Spin" the cycle of the pumping chain until infinite elements are created.

3. Dispose of them along the tail until the place σ is reached.

4. Repeat this procedure until there is no unprocessed x in notFin_Φ.

Each stage is carried out by the procedure PumpOneChain in Section 5.6, based on the information contained in a pumping chain $\mathcal{PC} = \langle \mathcal{D}, \eta, q_0, \mathsf{P} \rangle$ in PCS.

In the following, we try to explain why this procedure can be effectively executed.

Rather than directly modifying the assignment $M_{\zeta,\jmath}$, we manipulate the regions of its Venn diagram through their formative process ζ and the sequence PCS. It will turn out that at the end of these transformations the cardinalities of the finite Venn regions will be preserved, so that $\left|M_{\zeta,\jmath}\,x\right| = |M\,x| < \aleph_0$ will hold, for every $x \in \mathsf{Fin}_\Phi$, whereas we shall also have $|M\,x| \geqslant \aleph_0$, for every $x \in \mathsf{notFin}_\Phi$.

More precisely, $\mathcal{D}$ is a list of $N + 1$ sequences $\mathcal{D}_0, \mathcal{D}_1, \ldots, \mathcal{D}_N$ of action moves (and their preferred targets), which are to be inserted at positions $\eta_0, \eta_1, \ldots, \eta_N$, respectively, in the sequence $(A_\nu)_{\nu < \ell}$ of moves of ζ, where $\eta = \langle \eta_0, \eta_1, \ldots, \eta_N \rangle$. The initial sequence $\mathcal{D}_0$ contains a cycle (and possibly a tail) which has to be repeated ω times in order to generate infinite elements in the blocks associated with all of its places. These will be then propagated through the tail, if present, and, subsequently, through the remaining sequences $\mathcal{D}_1, \ldots, \mathcal{D}_N$. The pumping process starts from the place q_0 in $\mathcal{D}_0$. The conditions for a pumping chain in Definition 5.12 guarantee that the block corresponding to the place q_0 is non-null at time η_0 (condition (i)). It is also guaranteed that the cycling process and the subsequent propagation steps can take place (conditions (iii) and (iv)) and that infinite $\mathcal{P}$-nodes can distribute among their targets all the elements which are able to produce (condition (v)). To make the cycle spin, we need an element that plays the role of generator of the infinite. We borrow such an element element e from $q^{(\eta_0)}$. After a cycle, a newly generated element $e' \neq e$ is returned to $q^{(\eta_0)}$, to restore its cardinality. A distinction is maintained between 'old' and 'new' elements during the pumping process, by temporarily splitting each block into a **Minus** and a **Surplus** part, respectively for old and new elements. **Surplus** parts grow only during the pumping steps proper (with the only exception of $\mathcal{P}$-nodes at their grand-event moves during the imitation phase), whereas the **Minus** parts are synchronized with the corresponding blocks of the original formative process ζ and grow only during the imitation phases. The main purpose of **Minus** elements is to imitate the original process and to avoid that our pumping procedure disrupts properties that guarantee simulation of the original process. Thus it is rather natural that they do not play any role in the pumping procedure, which is totally devoted to the **Surplus** elements.

According to this, up to the instant η_0, imitation is carried out just by copying, whereas after the instant η_0, only cardinality is maintained between the original block and its corresponding **Minus** part, plus membership of $\subseteq$-maximal elements, to ensure $\in$-simulation.

More in detail, the above-mentioned manipulation of ζ is performed by the program ExtendFormativeProcess, with the help of the procedures PumpOneChain, which has inside procedures CopyInitialSegment, PumpCycle, Propagate, and Imitate. These are all reported in Section 5.6.

Roughly speaking the procedure PumpOneChain acts as follows:

- CopyInitialSegment copies faithfully the segment which precedes immediately the cycle;

- PumpCycle repeats the elements of the cycle ω-times, by creating an infinity of elements;

- Imitate imitates the segment between the cycle and the first cut of the tail

- Propagate propagates the infinite in the sense that it distributes the elements created by the cycle along the first cut of the tail;

- Again Imitate imitates the segment between the first and the second cut of the tail.

The above steps are repeated until the end of tail is reached. Subsequently, the remaining part of the original process is imitated.

ExtendFormativeProcess simply generalizes this procedure using several different cycles.

Observe that whenever a pumping procedure is applied the output process is actually shadow of the preceding one. In order to verify this fact, it is sufficient to apply standard arguments. By the Shadow Theorem 3.23 we obtain that the resulting partition simulates the starting one. Arguing as in Theorem 4.9 we get L-simulation.

An application of this procedure, when there is just one cycle, can be seen in the proof of Lemma 5.17, which is preparatory to showing the strong rank dichotomicity of MLSP.

The program ExtendFormativeProcess, which takes as input the initial finite formative process ζ and the sequence PCS of pumping chains in ζ, makes a call to the procedure PumpOneChain for each pumping chain in PCS in nondecreasing order of their starting times. Thus, it is convenient to assume that the sequence PCS is already so arranged.

As we shall see soon, the effect of a call to the procedure PumpOneChain, with input a formative process $\mathcal{H} = \left\langle (\Pi_\mu)_{\mu \leq \xi}, (\bullet), \mathcal{T}, \mathcal{R} \right\rangle$ and the first pumping chain $\mathcal{PC}$ of an ordered sequence PCS$'$ of pumping chains in $\mathcal{H}$, is the construction of another formative process $\widehat{\mathcal{H}} = \left\langle (\widehat{\Pi}_\mu)_{\mu \leq \widehat{\xi}}, [\bullet], \widehat{\mathcal{T}}, \widehat{\mathcal{R}} \right\rangle$ and of an increasing map $f : \{0, 1, \ldots, \xi\} \to \{0, 1, \ldots, \widehat{\xi}\}$ (to map the remaining pumping chains in PCS$'$ into the new formative process $\widehat{\mathcal{H}}$) such that

(i) the final partition $\widehat{\Pi}_{\widehat{\xi}}$ of $\widehat{\mathcal{H}}$ $(m-1)$-simulates the final partition Π_ξ of $\mathcal{H}$;

(ii) - $|\widehat{q}^{[\bullet]}| \geq |q^{(\bullet)}|$, for each place $q \in \Pi$,

   - $|\widehat{q}^{[\bullet]}| = |q^{(\bullet)}|$, for each place $q \in \mathcal{R}$,

   - $|\widehat{q}^{[\bullet]}| \geq \aleph_0$, for each place q in the pumping path of $\mathcal{PC}'$ (plus possibly other places in the $\mathcal{P}$-closed set of places of $\mathcal{PC}'$);

(iii) $f(\mathcal{PC}')$ is a pumping chain in $\widehat{\mathcal{H}}$, for each $\mathcal{PC}' = \left\langle \mathcal{D}', \langle \eta_0', \eta_1', \ldots, \eta_{N'}' \rangle, q_0', \mathsf{P}' \right\rangle$ in PCS$'$, where we are using the notation $f(\mathcal{PC}') := \left\langle \mathcal{D}', \langle f(\eta_0'), f(\eta_1'), \ldots, f(\eta_{N'}') \rangle, q_0', \mathsf{P}' \right\rangle$.

All the times we apply the pumping procedure, a transitive partition which $(m-1)$-simulates the preceding one is created. Therefore the first partition obviously $(m-1)$-simulates the last one. Thus, in particular, the inherited assignment satisfies all the literals in Φ^- and in Φ_{Fin}, plus, possibly, some of the literals in $\Phi_{\neg\mathsf{Fin}}$; then, by (i) and (ii) and by the considerations made in Section 5.1, the assignment $M_{\widehat{\mathcal{H}},\widehat{\chi}} x = \bigcup (\widehat{\chi}(x))^{[\bullet]}$ over $\mathrm{Vars}(\Phi)$ (where $\widehat{\chi}(x) = \{\widehat{q} \mid q \in \chi(x)\}$, for $x \in V_\Phi$) satisfies all the literals in Φ^- and in Φ_{Fin}, plus the same literals in $\Phi_{\neg\mathsf{Fin}}$ which were already satisfied by $M_{\mathcal{H},\chi}$, plus all the literals $\neg\mathrm{Finite}(x)$ in $\Phi_{\neg\mathsf{Fin}}$ such that $\chi(x)$ contains some place in the pumping path of $\mathcal{PC}'$. In addition, all the variables occurring in $\Phi_{\neg\mathsf{Fin}}$ have a pumping chain which can be pumped and the pumping procedure preserves pumping chains. The same holds, in particular, for those clauses in $\Phi_{\neg\mathsf{Fin}}$, which are not still satisfied by $M_{\widehat{\mathcal{H}},\widehat{\chi}}$. Then, thanks to (iii), the subsequent calls of the procedure PumpOneChain with input the last formative process and the next remaining pumping chain in PCS$'$, suitably mapped by f into the former, will allow one to satisfy also the residual clauses of type $\neg\mathrm{Finite}(x)$ in Φ.

Finally, assumptions (a) and (b) guarantee that the property to be a pumping chain is preserved in $M_{\zeta,\mathrm{J}}$ and PCS, as is immediate to check.

The mapping of the residual pumping chains in PCS′ into the current formative process is made at line 6 of the program ExtendFormativeProcess:

6. seqPumpChain := f[seqPumpChain];

If the variable seqPumpChain contains the residual list $\langle PC'', PC'', \ldots, PC^{(k)} \rangle$, then f[seqPumpChain] computes the list $\langle f(PC'), f(PC''), \ldots, f(PC^{(k)}) \rangle$.

We have already described how the procedure PumpOneChain works. Here we come into details following the code of the procedure. As said, it takes as input a (not necessarily finite) colored Π-process $\mathcal{H} = \left\langle (\Pi_\mu)_{\mu \leqslant \xi}, (\bullet), \mathcal{T}, \mathcal{R} \right\rangle$ and a pumping chain $PC = \left\langle \mathcal{D}, \eta, q_0, \mathsf{P} \right\rangle$ in it, where $\mathcal{D} = \langle \mathcal{D}_0, \ldots, \mathcal{D}_N \rangle$ and $\eta = \langle \eta_0, \ldots, \eta_N \rangle$, and it computes a new colored Π-process $\widehat{\mathcal{H}} = \left\langle (\widehat{\Pi}_\mu)_{\mu \leqslant \widehat{\xi}}, [\bullet], \widehat{\mathcal{T}}, \widehat{\mathcal{R}} \right\rangle$, along with an increasing map $f : \{0, 1, \ldots, \xi\} \to \{0, 1, \ldots, \widehat{\xi}\}$ (such that $f(\mu) = \mu$, for $0 \leqslant \mu < \eta_0$, and $f(\xi) = f(\widehat{\xi})$), which maps $\mathcal{H}$ into $\widehat{\mathcal{H}}$ in the sense that will be clarified below.

The new formative process $\widehat{\mathcal{H}}$ is computed in different stages. For each place $q \in \Pi$ and ordinal $\mu \leqslant \widehat{\xi}$, the set $\widehat{q}^{[\mu]}$ is temporarily maintained as a disjoint union of two sets, $\mathsf{Minus}(\widehat{q}^{[\mu]})$ and $\mathsf{Surplus}(\widehat{q}^{[\mu]})$, in such a way that all the sequences $\{\mathsf{Minus}(\widehat{q}^{[\mu]})\}_{\mu \leqslant \widehat{\xi}}$ and $\{\mathsf{Surplus}(\widehat{q}^{[\mu]})\}_{\mu \leqslant \widehat{\xi}}$, for $q \in \Pi$, are nondecreasing with respect to set inclusion.

Conceptually Surplus and Minus simply define a refining of the given partition. Only at the end of the computation of procedure PumpOneChain, precisely at lines 15–16, Minus-Surplus refining is canceled to be merged to form a 'regular' formative process.

The pumping process proper takes place in the Surplus sequences, while the Minus sequences are used to maintain a copy of the original blocks $q^{(\mu)}$. Actually, a quite faithful copy since the blocks could contain different elements which play the same role of the original ones. More precisely, the sequence of ordinals $\langle \eta_0, \ldots, \eta_N \rangle$ break the ordinal interval $[0 .. \xi]$ into the $N + 2$ subintervals $[\eta'_j .. \eta_{j+1}]$, for $j = -1, 0, \ldots, N$, where, for convenience, we have put $\eta_{N+1} := \xi$, $\eta'_{-1} := 0$, and $\eta'_i := \eta_i + 1$, for $i = 0, \ldots, N$. The extended formative process $(\widehat{\Pi}_\mu)_{\mu \leqslant \widehat{\xi}}$ is obtained by inserting $N + 1$ new 'pumping' segments in $(\Pi_\mu)_{\mu \leqslant \xi}$ (corresponding to the $N + 1$ paths $\mathcal{D}_0, \ldots, \mathcal{D}_N$ of PC) just after the positions $\eta_0, \ldots, \eta_N$, while suitably adapting the $N + 2$ original segments $(\Pi_\mu)_{\eta'_j \leqslant \mu \leqslant \eta_{j+1}}$, for $j = -1, 0, \ldots, N$. In what follows we shall go into the details of procedure PumpOneChain.

(I) initially, the segment $(\Pi_\mu)_{0 \leqslant \mu < \eta_0}$ is merely copied (procedure CopyInitialSegment), so that we have

$\mathsf{Minus}(\widehat{q}^{[\mu]}) = q^{(\mu)}$, $\mathsf{Surplus}(\widehat{q}^{[\mu]}) = \emptyset$, and $f(\mu) = \mu$, for $q \in \Pi$ and $0 \leqslant \mu < \eta_0$.

But then, in order to trigger the subsequent pumping phase, a generator element e is picked from $q^{(\eta_0)}$ (which is non-null by the definition of pumping chain); thus, we have

$\mathsf{Minus}(\widehat{q}^{[\eta_0]}) = q^{(\eta_0)} \setminus \{e\}$, $\mathsf{Surplus}(\widehat{q}^{[\eta_0]}) = \{e\}$

(lines 8–9 of procedure CopyInitialSegment).

From now on, the original segments cannot just be copied into the corresponding Minus blocks, but they have to be handled differently by procedure Imitate where new elements could be interchanged with the original ones.

(II) The first 'pumping' segment of length $\kappa_0 := \omega + k_0$, corresponding to the path $\mathcal{D}_0$, is inserted at position $\eta_0 + 1$. Such a segment is generated by cycling ω times around the pumping cycle of $\mathcal{D}_0$ (procedure PumpCycle) and then propagating the infinite via the tail $\overline{\mathcal{D}}_0$ of $\mathcal{D}_0$ (if present), where $k_0 = |(\overline{\mathcal{D}}_0)_{nodes}|$ (procedure Propagate). At the end, we have:

$$f(\eta_0) = \eta_0 + \kappa_0 ,$$

$$\left|\mathsf{Minus}(\widehat{q}^{[f(\eta_0)]})\right| = \left|q^{(\eta_0)}\right| , \quad \text{for } q \in \Pi , \text{ and}$$

$$\left|\mathsf{Surplus}(\widehat{p}^{[f(\eta_0)]})\right| \geqslant \aleph_0 , \quad \text{for } p \in (\mathcal{D}_0)_{places}.$$

Notice that after the first cycle through the pumping cycle of $\mathcal{PC}$ we have to complete the exchange of the generator element. For this purpose, a newly generated element e' is picked from $\mathsf{Surplus}(\widehat{q}_0^{[\eta_0+1]})$ and moved into $\mathsf{Minus}(\widehat{q}_0^{[\eta_0+1]})$, so as to restore the cardinality of the latter (lines 11–15 of procedure PumpCycle). The set $\mathsf{Surplus}(\widehat{q}_0^{[\eta_0+1]})$ is already large enough to fuel the pumping process, even after it loses the element e'.

(III) The subsequent original segments $\left(\Pi_\mu\right)_{\eta_j+1\leqslant\mu\leqslant\eta_{j+1}}$ are attached to the current formative process (procedure Imitate) at positions $f(\eta_j)+1$, respectively, for $j = 0, 1, \ldots, N$; however, as already remarked, they cannot be just copied. Since we mainly focus on structural properties of the Π-board, we are not interested in the nature of the elements, therefore it is enough that cardinalities are maintained (lines 2–5 of procedure Imitate), and that grand events are treated properly, also in the case of $\mathcal{P}$-nodes (lines 2 and 6–9 of procedure Imitate). Thus we have:

$$\left|\mathsf{Minus}(\widehat{q}^{[f(\eta_j)+i]})\right| = \left|q^{(\eta_j+i)}\right|, \quad \text{for } j = 0, 1, \ldots, N \text{ and } i = 1, 2, \ldots, \eta_{j+1} - \eta_j , \text{ and}$$

$$f(\eta_j + i) = f(\eta_j) + i, \quad \text{for } j = 0, 1, \ldots, N \text{ and } i = 1, 2, \ldots, \eta_{j+1} - \eta_j - 1 ,$$

as far as cardinalities and the map f are concerned. In particular, for $i = \eta_{j+1} - \eta_j$ we have $\left|\mathsf{Minus}(\widehat{q}^{[f(\eta_j)+\eta_{j+1}-\eta_j]})\right| = \left|q^{(\eta_{j+1})}\right|$, for $j = 0, 1, \ldots, N$.

As a special case (line 14 of procedure PumpOneChain), we also have:

$$f(\xi) = f(\eta_{N+1}) = f(\eta_N) + \eta_{N+1} - \eta_N = f(\eta_N) + \xi - \eta_N.$$

Thus, $\widehat{\xi} = f(\eta_N) + \xi - \eta_N$.

(IV) This is the stage of the procedure in which infinite elements are actually delivered to the "predestined" place. Indeed, the subsequent 'pumping' segments of length $\kappa_j := |(\mathcal{D}_j)_{nodes}|$ are inserted into the current formative process (procedure Propagate) at positions $f(\eta_{j-1}) + \eta_j - \eta_{j-1} + 1$, respectively, for $j = 1, 2, \ldots, N$. At the end of the j-th step, for $j = 1, 2, \ldots, N$, we have:

$$f(\eta_j) = f(\eta_{j-1}) + \eta_j - \eta_{j-1} + \kappa_j ,$$

$$\left|\mathsf{Minus}(\widehat{q}^{[f(\eta_j)]})\right| = \left|q^{(\eta_j)}\right| , \quad \text{for } q \in \Pi , \text{ and}$$

$$\left|\mathsf{Surplus}(\widehat{p}^{[f(\eta_j)]})\right| \geqslant \aleph_0 , \quad \text{for } p \in (\mathcal{D}_j)_{places}.$$

Each pumping step, in either procedure PumpOneChain or Propagate, is able to distribute infinite elements from a set of the form

$$\mathcal{P}^*\left((\widehat{B \setminus \{p\}})^{[\alpha]} \cup \left\{\mathsf{Minus}(\widehat{p}^{[\alpha]}), \mathsf{Surplus}(\widehat{p}^{[\alpha]})\right\}\right) \setminus \bigcup \widehat{\Pi}^{[\alpha]} ,$$

for some node $B \in (\mathcal{D}_j)_{nodes}$ and ordinal $f(\eta_j - 1) < \alpha < f(\eta_j)$ (with $j = 0, 1, \ldots, N$ and p the predecessor of B in $\mathcal{D}_j$), among one or two targets of B. This is possible since, as is not hard to prove recursively, the set $\mathsf{Surplus}(\hat{p}^{[\alpha]})$ contains infinitely many new elements that have not been used yet at step α, and, in addition, all the sets $\mathsf{Minus}(\hat{q}^{[\alpha]})$, for $q \in B$, are non-null, as shown in the lemma below. The latter fact implies that each set $\hat{q}^{[\alpha]} = \mathsf{Minus}(\hat{q}^{[\alpha]}) \cup \mathsf{Surplus}(\hat{q}^{[\alpha]})$ is also non-null, so that the set $\mathcal{P}^*\big((\widehat{B \setminus \{p\}})^{[\alpha]} \cup \{\mathsf{Minus}(\hat{p}^{[\alpha]}), \mathsf{Surplus}(\hat{p}^{[\alpha]})\}\big)$ is non-null as well: in fact, it contains infinitely new elements, thanks to the contribution of the set $\mathsf{Surplus}(\hat{p}^{[\alpha]})$.

Lemma 5.14. *Let $B \in (\mathcal{D}_j)_{nodes}$ and let α be an ordinal such that $f(\eta_j - 1) < \alpha \leqslant f(\eta_j)$, where $j = 0, 1, \ldots, N$. Then $\mathsf{Minus}(\hat{q}^{[\alpha]}) \neq \emptyset$, for every $q \in B$.*

Proof. Let $B \in (\mathcal{D}_j)_{nodes}$, with $j = 0, 1, \ldots, N$, and let $q \in B$. As shown above, we have

$$\left| q^{(\eta_j)} \right| = \left| \mathsf{Minus}(\hat{q}^{[f(\eta_j)]}) \right|$$
$$\left| q^{(\eta_j)} \right| = \left| \mathsf{Minus}(\hat{q}^{[f(\eta_{j-1}) + \eta_j - \eta_{j-1}]}) \right| = \left| \mathsf{Minus}(\hat{q}^{[f(\eta_j - 1) + 1]}) \right| ,$$

where the latter equality follows from

$$f(\eta_{j-1}) + \eta_j - \eta_{j-1} = f(\eta_{j-1}) + \eta_j - \eta_{j-1} - 1 + 1$$
$$= f(\eta_{j-1} + \eta_j - \eta_{j-1} - 1) + 1$$
$$= f(\eta_j - 1) + 1 .$$

But, for every α such that $f(\eta_j - 1) < \alpha \leqslant f(\eta_j)$, we have

$$\left| q^{(\eta_j)} \right| = \left| \mathsf{Minus}(\hat{q}^{[f(\eta_j - 1) + 1]}) \right| \leqslant \left| \mathsf{Minus}(\hat{q}^{[\alpha]}) \right| \leqslant \left| \mathsf{Minus}(\hat{q}^{[f(\eta_j)]}) \right| = \left| q^{(\eta_j)} \right| ,$$

which yields

$$\left| \mathsf{Minus}(\hat{q}^{[\alpha]}) \right| = \left| q^{(\eta_j)} \right| . \tag{5.4}$$

Since $\mathcal{P}^*(B^{(\eta_j)}) \neq \emptyset$ (from (iv) of Definition 5.12), it follows that $q^{(\eta_j)} \neq \emptyset$ for every $q \in B$. Thus, by (5.4), we conlude that $\mathsf{Minus}(\hat{q}^{[\alpha]}) \neq \emptyset$, for $f(\eta_j - 1) < \alpha \leqslant f(\eta_j)$, which is our thesis. $\qquad\square$

The procedure $\mathsf{PumpCycle}$ is simplified if compared with the original one, as appears in [CU14]. Indeed, in the old version one had to distinguish two cases in order to distribute elements along the cycle: in the first case, the pumping step distributes new elements among two targets, in the other the pumping step gives new elements to just one target (cf. line 8). Since the process can continue along the tail in any case, this exception has been avoided. In order to understand why one can argue in this way, see Exercise 5.4.

As already observed, the final partition $\widehat{\Pi}_{\bar{\xi}}$ of the enlarged formative process $\widehat{\mathcal{H}}$ (returned by procedure $\mathsf{PumpOneChain}$) $(m-1)$-simulates the final partition Π_ξ of the input process $\widehat{\mathcal{H}}$ (cf. Condition 1 at the end of Section 5.1). This is due to the fact that $\widehat{\mathcal{H}}$ is a shadow process of the previous one (readers are asked to prove this in the exercises). This in particular shows, as announced in the introduction of the present chapter, a completely different use of shadow technique. Indeed, the size shadow models increases at each application of the above mentioned procedure.

More in detail, $\widehat{\mathcal{H}}$ complies closely with the cardinality constraints of the process $\mathcal{H}$ (instruction 2 of procedure Imitate). In particular, in order to enforce $\in$-simulation and L-simulation, special care is taken of the $\subseteq$-maximal element $\bigcup \widehat{B}^{[\mu]}$ during the distribution of new elements, say, at step μ, relative to the node B. In fact, for the $\in$-simulation to hold, it is necessary that $\bigcup \widehat{B}^{[\widehat{\xi}]} \in \widehat{q}^{[\widehat{\xi}]}$ iff $\bigcup B^{(\xi)} \in q^{(\xi)}$, for $q \in \Pi$. In order to follow as faithfully as possible the former process, our choice is to dispose of the element $\bigcup \widehat{B}^{[\mu]}$ only when the move B at step μ corresponds to a grand move in the input formative process; see procedures PumpCycle (line 5), Propagate (line 4), and Imitate (line 2, cases (c) and (d)). Likewise, $\mathcal{P}$-simulation is enforced by instructions 2 (case (a)) and 6–9 of procedure Imitate.

From the above considerations, we can conclude that the decision procedure for MLSSPF (procedure SatifisfiabilityTestMLSSPF) is complete. By the soundness result, already argued in Section 5.5.1, we therefore have:

Theorem 5.15 (Decidability of MLSSPF). *The satisfiability problem for MLSSPF is decidable.*

In view of our remarks in (II)–(IV) above, it turns out that

$$f(\eta_j) = \begin{cases} \eta_0 + \kappa_0 & \text{if } j = 0 \\ f(\eta_{j-1}) + \eta_j - \eta_{j-1} + \kappa_j & \text{if } 1 \leqslant j \leqslant N \\ f(\eta_N) + \xi - \eta_N & \text{if } j = N+1 \,. \end{cases} \tag{5.5}$$

By iterating (5.5), we obtain:

$$\widehat{\xi} = f(\eta_{N+1}) = \eta_0 + \sum_{i=0}^{N} \kappa_i + \xi - \eta_0 = \eta_0 + \omega + k_0 + \sum_{i=1}^{N} \kappa_i + \xi - \eta_0 < \eta_0 + \omega \cdot 2 \,, \tag{5.6}$$

since $k_0 + \sum_{i=1}^{N} \kappa_i$ is just the (finite) length of $\mathcal{PC}$ (cf. Definition 5.12) and $\xi - \eta_0$ is finite, by the definition of pumping chains.

Let $\mathcal{H}_0 = \langle (\Pi_\mu)_{\mu \leqslant \xi_0}, (\bullet), \mathcal{T}, \mathcal{R} \rangle$ be a colored Π-process of finite length ξ_0 and let PCS $= \langle \mathcal{PC}', \mathcal{PC}'', \ldots, \mathcal{PC}^{(k)} \rangle$ be a sequence of pumping chains in it, ordered non-decreasingly by their starting times $\eta_0' \leqslant \eta_0'' \leqslant \ldots \leqslant \eta_0^{(k)}$. Since the pumping operation is carried out at most k times it is quite natural that the rank of the model cannot exceed rank $\omega \cdot (k+1)$. Even if this result is quite reasonable, in the following we sketch its proof.

By calling the program ExtendFormativeProcess with input $\mathcal{H}$ and PCS (thus, through a sequence of k calls to procedure PumpOneChain), a sequence $\mathcal{H}_1, \mathcal{H}_2, \ldots, \mathcal{H}_k$ of colored shadow processes, of lengths $\xi_1, \xi_2, \ldots, \xi_k$, respectively, is computed, along with a sequence $f_1, f_2, \ldots, f_k$ of indices correspondence maps, where, more precisely, $\mathcal{H}_i$ and $f_i : [0..\xi_{i-1}] \to [0..\xi_i]$ are the formative process and the indices correspondence map, respectively, computed after the i-th call to procedure PumpOneChain. Let us show, by inducting on i, that

$$\xi_i < \omega \cdot (i+1), \quad \text{for } i = 0, 1, \ldots, k \,. \tag{5.7}$$

The base case $\xi_0 < \omega$ holds by hypothesis. Next assume that $\xi_i < \omega \cdot (i+1)$ for $0 \leqslant i < k$.

From (5.6) we have

$$\xi_{i+1} < (f_i \circ f_{i-1} \circ \cdots \circ f_1)(\eta_0^{(i)}) + \omega \cdot 2 \leqslant \omega \cdot (i+2),$$

since $(f_i \circ f_{i-1} \circ \cdots \circ f_1)(\eta_0^{(i)}) \leqslant \xi_i < \omega \cdot (i+1)$ and therefore $(f_i \circ f_{i-1} \circ \cdots \circ f_1)(\eta_0^{(i)}) \leqslant \omega \cdot i + \alpha$, for some finite ordinal α, so that

$$(f_i \circ f_{i-1} \circ \cdots \circ f_1)(\eta_0^{(i)}) + \omega \cdot 2 \leqslant \omega \cdot i + \alpha + \omega \cdot 2 = \omega \cdot i + \omega \cdot 2 = \omega \cdot (i+2).$$

For $i = k$, (5.7) yields $\xi_k < \omega \cdot (k+1)$. Thus, we have

Lemma 5.16. *An* MLSSPF-*conjunction with k literals of the form $\neg Finite(x)$ is satisfiable if and only if it has a model of rank strictly less than $\omega \cdot (k+1)$.*

5.6   Extending a Formative Process with Respect to a Sequence of Pumping Chains

program ExtendFormativeProcess(inputFormProcess, seqPumpChain);

-- *formProcess is a finite greedy colored Π-process* $\mathcal{H} = \langle (\Pi_i)_{i \leqslant \ell}, (\bullet), \mathcal{T}, \mathcal{R} \rangle$ *and*

-- *seqPumpChain is a list of pumping chains in $\mathcal{H}$ ordered non-decreasingly by their starting*

-- *times.*

-- *Notation: for a non-null list $\mathcal{S}$, first$(\mathcal{E})$ and last$(\mathcal{S})$ are, respectively, the first and the last*

-- *element of $\mathcal{S}$, and dropFirst$(\mathcal{S})$ is the list obtained from $\mathcal{S}$ by dropping its first element.*

1. **global** Minus $:= \emptyset$,   Surplus $:= \emptyset$,   f $:= \emptyset$,   step $:= 0$;

-- *these are the global variables of the procedure; in particular, Minus and Surplus are*

-- *set-valued maps over $\Pi \times [0, \gamma]$, for some ordinal γ, which induce a Minus-Surplus*

-- *partitioning of the formative Π-process $(\Pi_\mu)_{\mu \leqslant \gamma} = (\{\text{Minus}(\widehat{q}^{[\mu]}) \cup \text{Surplus}(\widehat{q}^{[\mu]})\}_{q \in \Pi})_{\mu \leqslant \gamma}$*

-- *constructed by procedure PumpOneChain, f is an indices correspondence map which keeps*

-- *track of the unpumped segments of the colored Π-process formProcess, and step is the*

-- *current length of the formative process under construction.*

-- *The variables Minus, Surplus, f, and step are updated by the calls below to the*

-- *procedures CopyInitialSegment, PumpCycle, Propagate, and Imitate.*

2. **while** $|\text{seqPumpChain}| \geqslant 1$ **do**
3.   pumpChain $:=$ first(seqPumpChain);
4.   seqPumpChain $:=$ dropFirst(seqPumpChain);
5.   formProcess $:=$ PumpOneChain(formProcess, pumpChain);
6.   seqPumpChain $:=$ f[seqPumpChain];   -- *the residual pumping chains are mapped by* f
                              -- *into the current formative process* formProcess
7.   Minus $:= \emptyset$;   Surplus $:= \emptyset$;   f $:= \emptyset$;   step $:= 0$;   -- *global variables are reinitialized*
8. **end while**;
9. **return** formProcess;
end program;

procedure PumpOneChain(formProcess, pumpChain);

-- *formProcess is a colored Π-process* $\mathcal{H} = \langle (\Pi_\mu)_{\mu \leqslant \ell}, (\bullet), \mathcal{T}, \mathcal{R} \rangle$ *and* pumpChain *is a*

-- *pumping chain* $\langle \mathcal{D}, \eta, q_0, \mathsf{P} \rangle$ *in $\mathcal{H}$, where $\mathcal{D} = \langle \mathcal{D}_0, \mathcal{D}_1, \ldots, \mathcal{D}_N \rangle$ and $\eta = \langle \eta_0, \eta_1, \ldots, \eta_N \rangle$.*

-- *Notation: tail$(\mathcal{D}_0)$ is the tail of $\mathcal{D}_0$ (cf. Definition 5.4).*

1. CopyInitialSegment (formProcess, η_0, q_0);
2. PumpCycle (formProcess, pumpChain, η_0, q_0);
3. **if** $(\text{tail}(\mathcal{D}_0))_{nodes} \neq \emptyset$ **then**
4. $d_0 :=$ first(tail$(\mathcal{D}_0)$); $\overline{\mathcal{D}}_0 :=$ dropFirst(tail$(\mathcal{D}_0)$);
5.   Propagate (formProcess, $\overline{\mathcal{D}}_0$, η_0, d_0);   -- *this also sets* $f(\eta_0) :=$ step
6. **else**
7. $f(\eta_0) :=$ step;   -- *otherwise* $f(\eta_0) :=$ step *has to be set here*
8. **end if**;
9. **for** $i := 1$ **to** N **do**
10.   Imitate (formProcess, P, η_{i-1}, η_i);
11.   Propagate (formProcess, $\mathcal{D}_i$, η_i, last$(\mathcal{D}_{i-1})$);
12. **end for**;
13. Imitate (formProcess, P, η_N, ξ);
14. $f(\xi) :=$ step;   -- *because* $f(\xi)$ *has not been set by the previous call to procedure* Imitate
15. - let $q \mapsto \overline{q}^{[\circ]}$ be the map defined on Π by $\overline{q}^{[\circ]} :=$ Minus$(\widehat{q}^{[\text{step}]}) \cup$ Surplus$(\widehat{q}^{[\text{step}]})$;
16. - assign to the variable formProcess the colored Π-process $\langle (\overline{\Pi}_\mu)_{\mu \leqslant \text{step}}, [\circ], \mathcal{T}, \mathcal{R} \rangle$ such
    that $\overline{\Pi}_\mu = \{\overline{q}^{[\mu]} \mid q \in \Pi\}$, for each $\mu \leqslant$ step, where $\overline{q}^{[\mu]} =$ Minus$(\widehat{q}^{[\mu]}) \cup$ Surplus$(\widehat{q}^{[\mu]})$;

<pre>
17.   return formProcess;
end procedure;
</pre>

procedure CopyInitialSegment(formProcess, η, p);
    -- formProcess *is a colored* Π-*process* $\left\langle (\Pi_\mu)_{\mu \leq \xi}, (\bullet), \mathcal{T}, \mathcal{R} \right\rangle$, η *is a successor ordinal less*
    -- *than* ξ, *and* p *is a place in* Π *such that* $p^{(\eta)} \setminus p^{(\eta-1)} \neq \emptyset$;

<pre>
1.   for α := 0 to η do
2.   for q ∈ Π do
3.   Minus(q̂[α]) := q(α);   Surplus(q̂[α]) := ∅;
4.   end for;
5.   step := α;
6.   if α < η then f(α) := α; end if;
7.   end for;
8.   pick e ∈ p(η) \ p(η−1);   -- seed borrowing
9.   Minus(p̂[η]) := Minus(p̂[η]) \ {e};   Surplus(p̂[η]) := {e};
end procedure;
</pre>

procedure PumpCycle (formProcess, pumpChain, η, p);
    -- formProcess *is a colored* Π-*process* $\mathcal{H} = \left\langle (\Pi_\mu)_{\mu \leq \xi}, (\bullet), \mathcal{T}, \mathcal{R} \right\rangle$, pumpChain *is a pumping*
    -- *chain* $\left\langle \mathcal{D}, \eta, q_0, \mathrm{P} \right\rangle$ *in* $\mathcal{H}$, *where* $\mathcal{D} = \langle \mathcal{D}_0, \mathcal{D}_1, \ldots, \mathcal{D}_N \rangle$, *with pumping cycle* $\mathcal{C}$, p *is the*
    -- *place in* $\mathcal{C}$ *which has been used for seed extraction by procedure* CopyInitialSegment *at step* η.

    -- <u>*Notation:*</u> *for each place* $q \in \Pi$ *and node* $B \subseteq \Pi$, *we put*
    -- $\widehat{q}^{[\alpha]} := \mathrm{Minus}(\widehat{q}^{[\alpha]}) \cup \mathrm{Surplus}(\widehat{q}^{[\alpha]})$ *and* $\widehat{B}^{[\alpha]} := \{ \widehat{q}^{[\alpha]} \mid q \in \Pi \}$; *in addition,*
    -- $\mathrm{next}_\mathcal{C}(W)$ *denotes the successor of* W *in* $\mathcal{C}$, *and* $\mathrm{first}(\mathcal{D}_0)$ *is the first vertex of* $\mathcal{D}_0$.

<pre>
1.   p₀ := p;
2.   firstRound := true;
3.   for α ∈ [η, η + ω[do
4.   B := next_C(p);   r := next_C(B);
5.   - let {R, S} be a partition of the set
</pre>
$$\mathcal{P}^*\left((\widehat{B \setminus \{p\}})^{[\alpha]} \cup \left\{ \mathrm{Minus}(\widehat{p}^{[\alpha]}), \mathrm{Surplus}(\widehat{p}^{[\alpha]}) \right\} \right) \setminus \left(\bigcup \widehat{\Pi}^{[\alpha]} \cup \left\{ \bigcup \widehat{B}^{[\alpha]} \right\} \right);$$
<pre>
6.   Minus(r̂[α+1]) := Minus(r[α]);   Surplus(r̂[α+1]) := Surplus(r̂[α]) ∪ R ∪ S;
7.   remainingPlaces := Π \ {r};
8.   for q ∈ remainingPlaces do
9.   Minus(q̂[α+1]) := Minus(q̂[α]);   Surplus(q̂[α+1]) := Surplus(q̂[α]);
10.   end for;
11.   if firstRound and r = p₀ then   -- the first round has been just completed and
12.   - pick e ∈ Surplus(p̂₀[α+1]);   -- the 'seed' can be returned to Minus(p̂₀[α+1])
13.   Minus(p̂₀[α+1]) := Minus(p̂₀[α+1]) ∪ {e};   Surplus(p̂₀[α+1]) := Surplus(p̂₀[α+1]) \ {e};
14.   firstRound := false;
15.   end if;
16.   p := r;
17.   end for;
18.   step := η + ω;
19.   for q ∈ Π do   -- limit step
20.   Minus(q̂[η+ω]) := ⋃_{μ<η+ω} Minus(q̂[μ]);   Surplus(q̂[η+ω]) := ⋃_{μ<η+ω} Surplus(q̂[μ]);
21.   end for;
end procedure;
</pre>

procedure Propagate (formProcess , $\mathcal{E}$, η , p);

    -- formProcess is a colored Π-process $\mathcal{H} = \langle(\Pi_\mu)_{\mu \leqslant \xi}, (\bullet), \mathcal{T}, \mathcal{R}\rangle$, $\mathcal{E}$ is a finite simple subpath

    -- of a pumping chain of $\mathcal{H}$ starting with a node and ending with a place, p is the predecessor

    -- of $\mathcal{E}$, and η is an ordinal less than ξ;

1.   **for** $\alpha := \mathsf{step}$ **to** $\mathsf{step} + |(\mathcal{E})_{nodes}| - 1$ **do**
2. $B := \mathsf{next}_{\mathcal{E}}(p);$ $r := \mathsf{next}_{\mathcal{E}}(B);$
3. $\mathsf{Minus}(\widehat{r}^{[\alpha+1]}) := \mathsf{Minus}(\widehat{r}^{[\alpha]});$
4. $\mathsf{Surplus}(\widehat{r}^{[\alpha+1]}) := \mathsf{Surplus}(\widehat{r}^{[\alpha]}) \cup \mathcal{P}^*\Big((\widehat{B \setminus \{p\}})^{[\alpha]} \cup \big\{\mathsf{Minus}(\widehat{p}^{[\alpha]}), \mathsf{Surplus}(\widehat{p}^{[\alpha]})\big\}\Big)$
5. $\setminus \big(\bigcup\widehat{\Pi}^{[\alpha]} \cup \{\bigcup\widehat{B}^{[\alpha]}\}\big);$
6.   **for** $q \in \Pi \setminus \{r\}$ **do**
7. $\mathsf{Minus}(\widehat{q}^{[\alpha+1]}) := \mathsf{Minus}(\widehat{q}^{[\alpha]});$ $\mathsf{Surplus}(\widehat{q}^{[\alpha+1]}) := \mathsf{Surplus}(\widehat{q}^{[\alpha]});$
8.   **end for**;
9. $p := r;$
10. **end for**;
11. $\mathsf{step} := \mathsf{step} + |(\mathcal{E})_{nodes}|;$
12. $\mathsf{f}(\eta) := \mathsf{step};$
end procedure;

procedure Imitate (formProcess , P , η , θ);

    -- formProcess is a colored Π-process $\mathcal{H} = \langle(\Pi_\mu)_{\mu \leqslant \xi}, (\bullet), \mathcal{T}, \mathcal{R}\rangle$, P is a $\mathcal{P}$-closed set of

    -- places, and η and θ are ordinals less than ξ such that $\eta \leqslant \theta$, θ is a successor ordinal,

    -- and $\theta - \eta$ is finite;

1.   **for** $\alpha := \eta$ **to** $\theta - 1$ **do**
2.   - let $q \mapsto \nabla(\widehat{q}^{[\mathsf{step}]})$ be a set-valued map over Π such that
        (a) $\{\nabla(\widehat{q}^{[\mathsf{step}]}) \mid q \in \Pi\} \setminus \{\emptyset\}$ is a partition of a non-null subset of
            $\mathcal{P}^*(\mathsf{Minus}[\widehat{A}_\alpha^{[\mathsf{step}]}]) \setminus \widehat{\Pi}^{[\mathsf{step}]},$
        (b) $|\nabla(\widehat{p}^{[\mathsf{step}]})| = |\Delta^{(\alpha)}(p)|,$ for all $p \in \Pi,$
        (c) if either (c.i) $\mathsf{GE}(A_\alpha) \neq \alpha$ or (c.ii) $\mathsf{GE}(A_\alpha) = \alpha$ and $\bigcup\mathsf{Surplus}[\widehat{A}_\alpha^{[\mathsf{step}]}] = \emptyset,$
          then $\bigcup\mathsf{Minus}[\widehat{A}_\alpha^{[\mathsf{step}]}] \in \nabla(\widehat{p}^{[\mathsf{step}]}) \leftrightarrow \bigcup A_\alpha^{(\alpha)} \in \Delta^{(\alpha)}(p),$ for all $p \in \Pi;$
        (d) if $\mathsf{GE}(A_\alpha) = \alpha$ and $\bigcup\mathsf{Surplus}[\widehat{A}_\alpha^{[\mathsf{step}]}] \neq \emptyset,$ then
           $\bigcup\widehat{A}_\alpha^{[\mathsf{step}]} \in \nabla(\widehat{p}^{[\mathsf{step}]}) \leftrightarrow \bigcup A_\alpha^{(\alpha)} \in \Delta^{(\alpha)}(p),$ for all $p \in \Pi;$
3.   **for** $q \in \Pi$ **do**
4. $\mathsf{Minus}(\widehat{q}^{[\mathsf{step}+1]}) := \mathsf{Minus}(\widehat{q}^{[\mathsf{step}]}) \cup \nabla(\widehat{q}^{[\mathsf{step}]});$ $\mathsf{Surplus}(\widehat{q}^{[\mathsf{step}+1]}) := \mathsf{Surplus}(\widehat{q}^{[\mathsf{step}]});$
5.   **end for**;
6.   **if** $\mathsf{GE}(A_\alpha) = \alpha$ **and** $\bigcup\mathsf{Surplus}[\widehat{A}_\alpha^{[\mathsf{step}]}] \neq \emptyset$ **and** A_α is a $\mathcal{P}$-node **then**
7.   - let p be a local trash of A_α in P;
8. $\mathsf{Surplus}(\widehat{p}^{[\mathsf{step}+1]}) := \mathsf{Surplus}(\widehat{p}^{[\mathsf{step}+1]}) \cup \Big(\mathcal{P}^*(\widehat{A}_\alpha^{[\mathsf{step}]}) \setminus \widehat{\Pi}^{[\mathsf{step}]} \setminus \bigcup_{q \in \Pi} \nabla(\widehat{q}^{[\mathsf{step}]})\Big);$
9.   **end if**;
10. $\mathsf{step} := \mathsf{step} + 1;$
11.   **if** $\alpha < \theta - 1$ **then** $\mathsf{f}(\alpha + 1) := \mathsf{step};$ **end if**;   *-- to synchronize the new process with*
                                  -- the original one; $\mathsf{f}(\theta)$ will be set by the subsequent call
                                  -- to procedure Propagate, if any, or by procedure PumpOneChain;
12. **end for**;
end procedure;

5.7 Strong Rank Dichotomicity of MLSP

As a further application of the formative process technique, we prove that the theory MLSP is strongly rank dichotomic (see Definition 2.53).

Lemma 5.17. *Let Φ be a satisfiable* MLSP*-formula and M a model of Φ of finite rank. Let $\mathcal{H} = \left\langle \left(\Pi_i \right)_{i \leqslant \ell}, (\bullet), \mathcal{T} \right\rangle$ be a Π-process for the Venn $\mathcal{P}$-partition Π_ℓ of M, and assume that $\ell \geqslant 2^{2^{|Var(\Phi)|}}$. Then Φ admits an infinite model.*

Proof. It can be easily checked that Π_ℓ has a Π-board with a cycle $\mathcal{C}$ that is followed at least two times during the Π-process $\mathcal{H}$. Consider the following set of variables of Φ:

$$V_{\mathcal{C}} := \left\{ x \in \mathrm{Vars}(\Phi) \mid \exists \sigma \in \mathcal{C} \wedge \sigma \subseteq x \right\}.$$

Fix $x \in V_{\mathcal{C}}$. Define a new formula Φ' by putting:

$$\Phi' = \Phi \wedge \neg Finite(x).$$

Plainly, Φ' is a MLSSPF-formula.

In order to prove our thesis, it is sufficient to show that $\mathcal{C}$ is a pumping chain for σ. First we need to change our Π-process in a colored Π-process. With this aim in mind, we consider the following procedure:

```
procedure LocalTrash(formProcess, pumpCycle);
          -- formProcess is a Π-process H = ⟨(Πμ)μ≤ξ, (•), T⟩ and pumpCycle is a
          -- C = ⟨σ0, A0, σ1, A1 ··· σm, Am = A0⟩.
    1.    LT:= Cplaces;
    2.    - mark all P-nodes as "unchecked";
    3.    while there are unchecked P-nodes do
    4.        - pick an unchecked P-node B such that B ∩ LT ≠ ∅;
    5.        i := GE(B);
    6.        - pick σ such that Δ⁽ⁱ⁾(σᵢ₊₁) ≠ ∅;
    7.        LT := LT ∪ {σ};
    8.        - mark B as "checked";
    9.    end while;
    10.   return LT;
end procedure;
```

The above procedure takes as input a finite Π-process together with a simple cycle of the Π-board of the resulting transitive partition, and returns a collection of places LT.

Consider now the collection LT of places and the following colored Π-process $\mathcal{H} = \left\langle \left(\Pi_i \right)_{i \leqslant \ell}, (\bullet), \mathcal{T}, \mathcal{R} \right\rangle$, where $\mathcal{R} = \Pi \setminus$ LT.

By inspecting the procedure LocalTrash, we can deduce that LT $=$ P is a $\mathcal{P}$-closed set of places, i.e., LT $\subseteq \Pi \setminus \mathcal{R}$, and that every $\mathcal{P}$-node B intersecting P has a local trash in it. Indeed, pick a node B such that $B \cap$ LT $\neq \emptyset$. Then, by procedure LocalTrash, there is a $\sigma' \in$ LT which receives elements at GE(B). We show that such a σ' is inside LT, and that it is a local trash for B. The former assertion is clearly fulfilled. By arguing from the construction of procedure LocalTrash, the latter assertion depends on the fact that all the nodes distribute all their elements. Therefore, GE(A) $\geq$ GE(B) holds, for all A such that

$\sigma \in A$. Since the cycle is repeated two times, then $\mathsf{GE}(B) \geq i_0$ holds, for every $\mathcal{P}$-node B intersecting P, where i_0 is the step in which the cycle is completed.

We can summarize the above result just by saying that $\mathcal{PC} = \langle \mathcal{C}, i_0, \sigma, LT \rangle$ is a pumping chain for σ. $\square$

Lemma 5.17 implies at once the following result.

Corollary 5.18. *The theory* MLSP *is strongly rank dichotomic.*

EXERCISES

Exercise 5.1. *Observe that any Π-board has an empty node $\emptyset$ and a place $q \in \Pi$ such that $\mathcal{T}(\emptyset) = \{q\}$, which is, roughly speaking, the first place which receives an element. Since this happens for all Π-boards, we can call q the* ROOT *of Π. Assuming that Π has only three places $\{q_1, q_2, q_3\}$, calculate how many boards with root q_1 it is possible to construct. In addition, establish how many of them can be actually realized, in the sense that they are Π-boards of some transitive partition. Answer the same question also when $|\Pi| = 2$.*

Exercise 5.2. *Consider all Π-boards with root q (see the preceding exercise) such that $q \in \mathcal{T}(\{q\})$. Find out which of them exist and why.*

Exercise 5.3. *Prove that an application of procedure* ExtendFormativeProcess *creates a shadow model.*

Exercise 5.4. *Consider the following pumping cycle*

$$\begin{pmatrix} q_1^{(\mu)} \\ q_2^{(\mu)} \end{pmatrix}_{\mu \leq \omega} = \begin{pmatrix} \emptyset & \{\emptyset\} & \{\emptyset\} & \{\emptyset\}, \{\{\{\emptyset\}\}\} & \cdots \\ \emptyset & \emptyset & \{\{\emptyset\}\} & \{\{\emptyset\}\} & \cdots \end{pmatrix},$$

with history and trace

$$\begin{pmatrix} A_\mu \\ T_\mu \end{pmatrix}_{\mu < \omega} = \begin{pmatrix} \emptyset & \{q_1\} & \{q_2\} & \cdots \\ \{q_1\} & \{q_2\} & \{q_1\} & \cdots \end{pmatrix}.$$

Assume one continues with the following two different syllogistic boards, respectively,

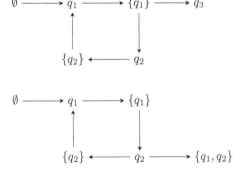

Say why, in both cases, it is possible to follow the different tails after the limit step, or in other terms, say why either $\mathcal{P}^*(\{q_1^{(\omega)}\}) \setminus \bigcup \Pi^{(\alpha)}$ or $\mathcal{P}^*(\{q_1^{(\omega)}, q_2^{(\omega)}\}) \setminus \bigcup \overline{\Pi}^{(\alpha)}$ is not empty, where $\Pi := \{q_1, q_2, q_3\}$, $\overline{\Pi} := \{q_1, q_2\}$, and $\alpha < \omega$.

References

[Acz88]   P. Aczel. *Non-well-founded Sets*, vol. 14 of *CSLI Lecture Notes*. CSLI, Stanford, 1988.

[AP88]   P. Atzeni and D.S. Park. Set containment inference and syllogisms. *Theoretical Computer Science*, 62:39–65, 1988.

[Can87]   D. Cantone. *A decision procedure for a class of unquantified formulae of set theory involving the powerset and singleton operators*. PhD thesis, New York University - GSAS, Courant Institute of Mathematical Sciences, February 1987.

[Can91]   D. Cantone. Decision procedures for elementary sublanguages of set theory. X. Multilevel syllogistic extended by the singleton and powerset operators. *Journal of Automated Reasoning*, 7(2):193–230, 1991.

[CC91]   D. Cantone and V. Cutello. Decision procedures for elementary sublanguages of set theory. XVI. Multilevel syllogistic extended by singleton, rank comparison and unary intersection. *Bulletin of EATCS*, 39:139–148, 1989.

[CCP89]   D. Cantone, V. Cutello, and A. Policriti. Set-theoretic reductions of Hilbert's tenth problem. In E. Börger, H. Kleine Büning, and M. M. Richter, editors, *Proceedings of 3rd Workshop Computer Science Logic - CSL '89 (Kaiserslautern 1989)*, volume 440, pages 65–75, Berlin, 1990. Springer-Verlag.

[CF95]   D. Cantone and A. Ferro. Techniques of computable set theory with applications to proof verification. *Comm. Pure Appl. Math.*, 48(9-10):1–45, 1995. Special issue in honor of J.T. Schwartz.

[CFO89]   D. Cantone, A. Ferro, and E.G. Omodeo. *Computable Set Theory*, vol. 6 International Series of Monographs on Computer Science. Clarendon Press, Oxford, UK, 1989.

[CFS85]   D. Cantone, A. Ferro, and J. T. Schwartz. Decision procedures for elementary sublanguages of set theory. VI. Multilevel syllogistic extended by the powerset operator. *Comm. Pure Appl. Math.*, XXXVIII(1):549–571, 1985.

[CFS87]   D. Cantone, A. Ferro, and J.T. Schwartz. Decision procedures for elementary sublanguages of set theory. V: Multilevel syllogistic extended by the general union operator. *Journal of Computer and Systems Sciences*, 34 1:1–18, 1987.

© Springer International Publishing AG, part of Springer Nature 2018
D. Cantone and P. Ursino, *An Introduction to the Technique of Formative Processes in Set Theory*, https://doi.org/10.1007/978-3-319-74778-1

[CGGW16] D. Cantone, A. Giarlotta, S. Greco, and S. Watson. (m, n)-rationalizable choices. *Journal of Mathematical Psychology* 73: 12–27, 2016.

[CGW17] D. Cantone, A. Giarlotta, and S. Watson. The satisfiability problem for Boolean set theory with a choice correspondence. In P. Bouyer, A. Orlandini and P. San Pietro: Proc. Eighth International Symposium on *Games, Automata, Logics and Formal Verification* (GandALF 2017), Roma, Italy, 20-22 September 2017, Electronic Proceedings in Theoretical Computer Science 256, pp. 61–75.

[COP01] D. Cantone, E.G. Omodeo, and A. Policriti. *Set Theory for Computing - From decision procedures to declarative programming with sets.* Monographs in Computer Science. Springer-Verlag, New York, 2001.

[COSU03] D. Cantone, E. G. Omodeo, J. T. Schwartz, and P. Ursino. Notes from the logbook of a proof-checker's project. In N. Dershowitz, editor, *Verification: Theory and Practice (Essays Dedicated to Zohar Manna on the Occasion of His 64th Birthday)*, vol. 2772 of *Lecture Notes in Computer Science*, pp. 182–207, Springer-Verlag, Berlin, 2003.

[COU99a] D. Cantone, E. G. Omodeo, and P. Ursino. Formative processes with applications to the decision problem in set theory. *Technical Report IASI-CNR*, 518, December 1999.

[COU99b] D. Cantone, E. G. Omodeo, and P. Ursino. Transitive Venn diagrams with applications to the decision problem in set theory. *Proc. of the AGP99 Joint Conference on Declarative Programming, L'Aquila, Italy*, September 1999.

[COU02] D. Cantone, E. G. Omodeo, and P. Ursino. Formative processes with applications to the decision problem in set theory: I. Powerset and singleton operators. *Information and Computation*, 172(2):165–201, January 2002.

[CU97] D. Cantone and P. Ursino. A unifying approach to computable set theory. In *Proceedings of Logic Colloquium 1997*, pp. 82–83, Leeds, UK, July 6-13, 1997.

[CU14] D. Cantone and P. Ursino. Formative processes with applications to the decision problem in set theory: II. Powerset and singleton operators, finiteness predicate. *Inf. Comput.*, 237:215–242, 2014.

[CZ99] D. Cantone and C.G. Zarba. A tableau-based decision procedure for a fragment of set theory involving a restricted form of quantification. In N. Murray, editor, *Proc. of the International Conference on Theorem Proving with Analytic Tableaux and Related Methods (TABLEAUX'99)*, LNCS 1617 (LNAI), pp. 97–112, Springer-Verlag, Berlin, 1999.

[CZ00] D. Cantone and C.G. Zarba. A new fast tableau-based decision procedure for an unquantified fragment of set theory. In R. Caferra and G. Salzer, editors, *Automated Deduction in Classical and Non-classical Logics*, LNCS 1761 (LNAI), pp. 126–136, Springer-Verlag, Berlin, 2000.

[CRF15]   M. Cristiá, G. Rossi, Claudia S. Frydman. Adding partial functions to Constraint Logic Programming with sets. *Theory and Practice of Logic Programming*, 15(4-5):651–665, 2015.

[Dav73]   M. Davis. Hilbert's tenth problem is unsolvable. *Amer. Mathematical Monthly*, 80:233–269, 1973.

[DMP95]   G. D'Agostino, A. Montanari, and A. Policriti. A set-theoretic translation method for polymodal logics. *Journal of Automated Reasoning*, 3(15):317–337, 1995.

[DPPR98]   A. Dovier, C. Piazza, E. Pontelli, and G.-F. Rossi. On the representation and management of finite sets in CLP-languages. In J. Jaffar, editor, *Proc. of 1998 Joint International Conference and Symposium on Logic Programming*, pp. 40–54, Manchester, UK, The MIT Press, 1998.

[FOS80]   A. Ferro, E.G. Omodeo, and J.T. Schwartz. Decision procedures for elementary sublanguages of set theory. I: Multilevel syllogistic and some extensions. *Comm. Pure Appl. Math.*, 33:599–608, 1980.

[GJ90]   M. R. Garey and D. S. Johnson. *Computers and Intractability; A Guide to the Theory of NP-Completeness*. W. H. Freeman, New York, USA, 1990.

[GW14]   A. Giarlotta and S. Watson. The pseudo-transitivity of preference relations: strict and weak (m, n)-Ferrers properties. *Journal of Mathematical Psychology* 58: 45–54, 2014.

[Jec78]   T.J. Jech. *Set Theory*. Academic Press, New York, 1978.

[KP95]   J.-P. Keller and R. Paige. Program derivation with verified transformations - A case study. *Comm. Pure Appl. Math.*, 48(9-10):1053–1113, 1995. Special issue in honor of J.T. Schwartz.

[Hil00]   David Hilbert, Mathematichse Probleme, Vortrag, gehalten auf dem internationalen Mathematiker-Kongress zu Paris 1900. *Nachrichten Akad. Wiss. Göttingen, Math.-Phys. Kl.* (1900) 253-297. English translation: *Bull. Amer. Math. Soc.*, 8 (1901-1902) 437-479.

[Mat93]   Y.V. Matiyasevich. Enumerable sets are Diophantine (Russian), *Dokl. Akad. Nauk SSSR*, 191:279–282, 1970. Improved English translation: *Soviet Math. Doklady*, 11:354-357, 1970.

[Mat93]   Y.V. Matiyasevich. *Hilbert's Tenth Problem*. The MIT Press, Cambridge, MA, 1993.

[OCPS06]   E. G. Omodeo, D. Cantone, A. Policriti, and J. T. Schwartz. A Computerized Referee. In M. Schaerf and O. Stock, editors, *Reasoning, Action and Interaction in AI Theories and Systems – Essays dedicated to Luigia Carlucci Aiello*, vol. 4155 of *Lecture Notes in Artificial Intelligence*, pp. 117–139. Springer Berlin/Heidelberg, 2006.

[OP10]   E. G. Omodeo and A. Policriti. The Bernays-Schönfinkel-Ramsey class for set
            theory: Semidecidability. *J. Symbolic Logic*, 75(2):459–480, 2010.

[OP12]   E. G. Omodeo and A. Policriti. The Bernays-Schönfinkel-Ramsey class for set
            theory: Decidability. *J. Symbolic Logic*, 77(3):896–918, 2012.

[OP16]   E. G. Omodeo and A. Policriti (Eds.). *Martin Davis on Computability, Com-
            putational Logic, and Mathematical Foundations.* vol. 10 of *Outstanding Con-
            tributions to Logic.* Springer International Publishing, 2016.

[OS02]   E. G. Omodeo and J. T. Schwartz. A 'Theory' mechanism for a proof-verifier
            based on first-order set theory. In A. Kakas and F. Sadri, editors, *Computa-
            tional Logic: Logic Programming and Beyond – Essays in honour of Bob Kowal-
            ski, Part II*, vol. 2048 of *Lecture Notes in Artificial Intelligence*, pp. 214–230.
            Springer-Verlag, Berlin, 2002.

[PP88]   F. Parlamento and A. Policriti. The logically simplest form of the infinity
            axiom. *Proceedings of the American Mathematical Society*, 103(1):274–276,
            1988.

[PPR97]   F. Parlamento, A. Policriti, and K. P. S. B. Rao. Witnessing differences
            without redundancies. *Proceedings of the American Mathematical Society*,
            125(2):587–594, 1997.

[Rog67]   H. Rogers. *Theory of recursive functions and effective computation*, McGraw-
            Hill, 1967.

[RB15]   G. Rossi and F. Bergenti, Nondeterministic Programming in Java with JSetL.
            Fundam. Inform., 140(3-4):393–412, 2015.

[Sam38]   P. Samuelson. A note on the pure theory of consumer's behavior. *Economica*
            5: 61–71, 1938.

[SCO11]   J. T. Schwartz, D. Cantone, and E. G. Omodeo. *Computational logic and set
            theory: Applying formalized logic to analysis.* Springer-Verlag, 2011. Foreword
            by M. Davis.

[SDDS86]   J. T. Schwartz, R. K. B. Dewar, E. Dubinsky, and E. Schonberg. *Programming
            with sets.* Texts and Monographs in Computer Scienc, Springer-Verlag, 1986.

[Sip12]   M. Sipser. *Introduction to the Theory of Computation*, Cengage Learning, 3rd
            edition, 2012.

[Urs05]   P. Ursino. A generalized small model property for languages which force the
            infinity. *Matematiche (Catania)*, LX(I):93–119, 2005.

[Zer77]   E. Zermelo. Untersuchungen über die Grundlagen der Mengenlehre I. In J. van
            Heijenoort, editor, *From Frege to Gödel - A Source Book in Mathematical Logic,
            1879-1931*, pp. 199–215. Harvard University Press, 1977. (3$^{\rm rd}$ Printing).

Acronyms

Operators and Common Symbols

© Springer International Publishing AG, part of Springer Nature 2018

D. Cantone and P. Ursino, *An Introduction to the Technique of Formative Processes in Set Theory*, https://doi.org/10.1007/978-3-319-74778-1

Index

Ackermann encoding, **48**, 58
action, 76, **76**
admissible flow from a node, 66
admissible flow-distribution step, 65, 66
Axiom
    Choice, **3**, 8
    Extensionality, **3**, 6
    Foundation, *see* Axiom of Regularity
    Regularity, **3**, 6, 8
    Specification, **3**, 6

bijection, **4**, 124
binary relation
    antisymmetric, **6**
    reflexive, **6**
    symmetric, **6**
    total, **6**
    transitive, **6**
bipartite directed graph, 60, 65
block, *see* partition
BST (Boolean Set Theory), 21, **39**, 40, 59
    decidability, 40
    normalized BST-conjunction, 40, 42–44
    procedure BST-SatTest, 40
BST (Boolean Set Theory), 58

cardinal, **8**, 9
cardinality comparison, **50**, 53
Cartesian product, **24**
    unordered C. p., **24**
choice, 19
    c. domain, 19
    rationalizable c., 19
Computable Set Theory, 21, 28
contractive map, **18**, 19

Davis, Martin, 53
decision
    d. problem, 26
    d. procedure, 26
Diophantine equation, 53
distinguishability, 42
    weakly d., 42
distribution edge, 72

edge-activation
    event, 67, **67**, 68, 82, 87, 115
    map, **85**, 87, 91, 92
equisatisfiability, **25**
equivalence relation, **6**, 12, 15, 20, 91
expressiveness
    e. of operations, **45**
    e. of relations, **45**

faithfulness condition, **48**, 58
finite part of a theory, 55, **55**
    hereditarily f. p. of a t., 55, **55**
flow network, 60, 65
formative process, 60, 62, 63, 65–68, 70–72, 76, **76**, 77–79, 81–84, 91, 92, 95–97, 99, 101–103, 105–107, 111, 112, 115, 118
    greedy f. p., 63–65, **77**, 78–80, 92, 97, 98, 100, 101, 103
    history of a f. p., 77
    special event, 60
    trace of a f. p., 66, 77
    weak f. p., 63, **77**, 78, 79, 99, 106, 107
function, **4**
    bijective, **4**, 33, 50, 56, 58
    composition, **4**
    domain, **4**

Printed in the United States
By Bookmasters